Information MANAGEMENT

Titles in the

TEAM LEADER DEVELOPMENT SERIES

Information Management

Sally Palmer

Head of Stockport College of Further and Higher Education Business School.

Margaret Weaver

Lecturer in Business Studies, Stockport College of Further and Higher Education. Fellow of Association of Chartered and Certified Accountants.

Information Management unlocks all the essential communication skills for today's supervisors and team leaders. It includes:

- presentation skills from OHPs to video conferencing.
- the supervisor's role in team briefings and feedback.
- a chapter on how to write a management project.

0 7506 3862 1: paperback: June 1998

People and Self Management

Sally Palmer

Head of Stockport College of Further and Higher Education Business School.

People and Self Management leads the reader through all the skills needed for today's supervisor/team leader, including:

- how to assess and improve your workplace performance.
- the essential skills of effective self management.
- the management of change.

0 7506 3861 3: paperback: June 1998

Resources Management

Margaret Weaver

Lecturer in Business Studies, Stockport College of Further and Higher Education. Fellow of Association of Chartered and Certified Accountants.

Resources Management is the absolute guide to all areas of resource control. It includes:

- thorough coverage of all areas of resource control for supervisors.
- clear explanations of theories and techniques of control.
- practical exercises to reinforce skills and knowledge.
- application of theory to the work-based problems facing today's managers.

0 7506 3863 X : paperback: June 1998

Activities Management

Cathy Lake

Freelance management writer.

Activities Management is a comprehensive guide to running a smooth and successful operation. It includes:

- practical help on how to plan and manage work.
- health and safety in the workplace.
- environmental considerations that today's supervisor needs to know.
- how to become a quality focussed organization

0 7506 4042 1: paperback: October 1998

Information MANAGEMENT

Team Leader Development Series

Sally Palmer
Margaret Weaver

OXFORD BOSTON JOHANNESBURG MELBOURNE NEW DELHI SINGAPORE

Butterworth-Heinemann
Linacre House, Jordan Hill, Oxford OX2 8DP
225 Wildwood Avenue, Woburn, MA 01801-2041
A division of Reed Educational and Professional Publishing Ltd

A member of the Reed Elsevier plc group

First published 1998

British Library Cataloguing in Publication Data
A catalogue record for this book is available from the British Library.

ISBN 0 7506 3862 1

Composition by Genesis Typesetting, Rochester, Kent
Printed and bound in Great Britain

Contents

3 Gathering data 22

4 Inputting and processing data to produce information 36

5 The use of information systems 55

9 Communication in organizations and workteams 113

10 Analysis and interpretation of information 129

Introduction

Introduction

There are four books in the Team Leader Development Series, *People and Self Managment*, *Information Managment*, *Resources Management* and *Activities Managment*, covering key topics from the four principal roles of management. The series has been designed to provide you with the knowledge and skills needed to carry out the role of team leader. The actual name of the job role of a team leader will vary from organization to organization. In your organization, the job role might be called any of the following:

- team leader
- supervisor
- first line manager
- section leader
- junior manager
- chargehand
- foreman
- assistant manager
- administrator.

If you work in the services or a hospital, team leaders may be called by another name not on the above list. However, in this series 'team leader' has been used throughout to describe the job role.

Who the series is intended for

If you have line-management responsibility for people within your organization, or you are hoping to progress to a position in which you will have this responsibility, then this series is for you. You may have been recently promoted into a team leader position or you may have been a team leader for some time. The series is relevant for you whether you work in a small organization or a large organization, whether you work in the public sector, private sector or voluntary sector. The books are designed to provide you with practical help which will enable you to perform better at work and to provide support to a range of programmes of study which have been designed specifically for team leaders.

Related programmes of study

There are a number of management qualifications that have been designed for team leaders. The titles in this series have been structured around the four key roles of management: Managing People, Managing Activities, Managing Resources and Managing Information. The content of each title has been developed in accordance with all the main qualifications in this area. Your tutor, manager or trainer will help you design a programme of study for your particular qualification route. Further details about each syllabus can be found in the tutor supplement that accompanies this textbook.

Information Management covers the core topics in this key role of management detailed in the programmes of study from the National Examining Board of Supervision and Management, the Institute of Supervisory Management, Edexel and the Institute of Management who all award qualifications in Supervisory Management. The Team Leader Development Series has also been devised to provide material that is relevant for those who are working towards a NVQ or SVQ at level 3 in management. The national management standards at this level cover the full range of general management activities which all managers working in a team leader position are expected to carry out. The Team Leader Development Series covers all the core topics involved with the activities defined in each of the key roles of management listed above. Your tutor will have full details about the national standards.

The content of *Information Management* covers the essential underpinning knowledge for the following mandatory unit:

D1 Manage information for action

This unit of competence consists of three elements:

D1.1 Gather required information
D1.2 Inform and advise others
D1.3 Hold meetings

The work-based assignments, which can be used to gather evidence for your portfolio, are mapped to the relevant elements of competence so that you can see which elements you are working towards.

As part of your work towards a vocational qualification in management at level 3, you also have to demonstrate that you have developed a number of personal competencies (in other words, skills and attitudes) that will enable you to apply your knowledge and understanding to a range of different situations at work. You will cover the range of personal competencies in many aspects of your work. This book will be particularly helpful in providing support for the following personal competencies:

- communicating
- searching for information
- thinking and taking decisions

Synopsis of *Information Management*

This textbook looks at two key areas: information and communication. The first section of the book covers what information is, why we need information, using information for decision making, gathering data and inputting and processing data to produce information. The important subjects of information technology and information systems are also covered in the first section of the book. Having completed this section you will have a good understanding of the importance of information and so will move on to look at how to communicate information. All aspects of communication are examined, such as oral communication, written communication including writing letters, memos and reports. Organizational communication and communicating with your team are also dealt with. The analysis and interpretation of information are covered and this chapter includes how to present information in the form of graphs and charts so that it is easier for the reader to understand. The final chapter in this book is entitled 'Undertaking a project and writing the project report'. Simple guidelines are provided to assist you with producing a project report. It is likely that you will have to undertake a project and write a project report if you are studying for a management qualification.

Learning structure

Each chapter begins with **Learning objectives**, a list of statements which say what you will be able to do, after you have worked through the chapter. This is followed by the 'Introduction', a few lines which introduce the material that is covered in the chapter.

There are several **Activities** in each chapter. You will find the answers at the end of the book.

There are also **Investigates** in each chapter. These are related to something which has been covered in the text. The suggestion is that you investigate the matter that has just been covered in your own organization. It is important, that you understand what you have learned, but also that you can relate what you have learned to your own organization.

Each chapter has a **Summary**. The summary recaps the main points that have been covered in the chapter. It rounds off the knowledge and skill areas that have been covered in the main body of the chapter, before the text moves into a range of tasks that you can complete to consolidate your learning.

There is a set of **Review and discussion questions** following the summary. You answer these after you have worked through the chapter to check whether you have understood and remembered the information that you have just read. Answers and guidelines to these questions can be found in the tutor resource material.

You are provided with an opportunity to deal with the issues raised in the chapter that you have just read by analysing the **Case study**. The case study is scenario based in the workplace and a chance to 'practise' how you might deal with a situation at work.

There is a **Work-based assignment** at the end of each chapter. These have been designed, so that if you complete the assignment, you will be able to apply the knowledge and skills that you have covered in the chapter in the workplace.

The relevant elements of competence are shown in the portfolio icon where applicable. These will be of use to you if you are studying towards an S/NVQ at Level 3 in management.

1 The need for information

Learning objectives

On completion of this chapter you will be able to:

- describe what is meant by 'information'
- explain why information needs to be managed
- identify different types of information
- explain the difference between quantitative and qualitative information
- describe the characteristics of good information.

Introduction

Information is needed for many purposes. It provides the basic material for many of the decisions which organizations, and the individuals within them, need to make. Without the right type of information, effectively communicated, making decisions is at best haphazard and at worst positively dangerous. This chapter will introduce you to the various types of information, and will enable you to assess the suitability of information for its purpose.

What is information?

Information is anything which is meaningful and useful to the person receiving it. It follows that if it is not meaningful and useful, then it is not information. In other words, it needs to *inform* the receiver of something which will enable him or her to perform their duties.

A very simple example of information used by most people is a telephone number. If you need to call someone on the telephone, you cannot do it without having the number! At the other end of the scale, a board of directors considering a proposal to extend the company's operations outside Britain will need information on the company's financial situation, world markets and economies, interest rates, foreign currencies, international trade laws, and so on.

Who uses information?

Well, the answer to this question should be obvious! Everyone uses information. It is used in routine, day-to-day tasks, infrequent or irregular activities and even in one-off situations. Individuals need information, as do groups and teams of people.

Users of information range from the most junior member of an organization to the most senior member, and even those outside it.

Internal users

Internal users include employees at all levels. They require information to enable them to do their jobs, to calculate their pay, to determine their holiday entitlement, to establish the stability of the company or to negotiate a pay rise.

Some items of information might be useful to several people (like the telephone number of your Head Office), while other items are useful to only one or two. It follows that information needs to be accessible to those who need it.

A team leader will use information for:

- recording
- planning
- monitoring
- controlling
- communicating

the activities of the section or group which they are managing.

External users

These fall into five broad groups:

1. *Customers*. They need information on the products or services available to them, such as product descriptions, the availability of stock, prices, delivery dates, payment periods and customer service standards.
2. *Suppliers*. They need information on the products and services too, but also on the ability of the organization to pay their bills on time.

3 *Owners*. They need access to almost any kind of information. If they are shareholders in a company, they will need information on the performance of the company, the share of profits they might receive and the future plans of the organization.
4 *The government*. Various government departments require information, to calculate taxes, analyse labour statistics, monitor pollution levels, etc.
5 *Society*. There are increasing demands from society in general to know more about the activities, both financial and non-financial, of organizations.

Activity 1

List the people who use the information provided by your department at work. Categorize them into *internal* and *external* users.

Why does information need to be managed?

Information is very important. Providing it is expensive and often time-consuming. You can probably recall several occasions when you have needed information for a particular purpose and could not locate it. Perhaps after a short search in vain, you have had to ask someone to help you - and then two people were involved in the same search. It might even be that you have had to manage without the information - and found your task much more difficult. Or perhaps you found what you *thought* was the right information and used it even though you were uncertain as to its reliability. Taken a step further, you might have made a wrong decision because of the lack of that piece of information, costing you and your organization time and money.

Equally important is the need to ensure that we do not have *too much* information for our needs, or time is wasted in sifting through useless items to locate the required piece of information.

Organizations cannot afford to misuse information in this way. Today's world is fast-moving and highly competitive, and opportunities need to be taken as they arise. Only by having the appropriate information available can they hope to succeed. Otherwise, they can be sure that others will come along and grasp those opportunities instead. It is to do with 'survival of the fittest' in many ways.

Most organizations nowadays recognize that information is an important *resource* - just as important as other resources such as materials, equipment, money, people and time. They have developed systems to manage those resources and are now also realizing the need to have systems to manage information.

In addition, today's world is highly computerized, and later in this book you will look at the use of computers in providing information. But computer technology is expensive to acquire and maintain, and it is important that the best possible use is made of it.

Activity 2

List three reasons why effective management of information is important.

See Feedback section for answer to this activity.

Types of information

Information can be categorized in many different ways, according to:

- its *timing* or *frequency*
- the *degree of formality* required
- its *source*
- its *method of communication*
- the extent to which it can be *quantified* (i.e. measured).

These categories are not exhaustive. For example, you could add 'the level of precision required'.

Timing and frequency of information

Much information is required on a regular and routine basis, hourly, daily, weekly, etc. Examples include:

- the price of a pint of beer in a pub
- the appointments schedule for a doctor's surgery
- the number of seats available on the next flight to Paris
- the quantity of goods on a supermarket shelf.

Information like this needs to be provided promptly, otherwise it is quickly out of date and therefore is not useful.

Imagine standing at the bar of the pub to be told that the price of beer is listed on a sheet of paper which no-one can find. Or arriving at a major supermarket to find there is no cat food on the shelves. An organization with such poor information would quickly lose customers, and possibly profit.

A system needs to be in place which gives this kind of information immediately, and which is very accurate.

Other information is needed less frequently, maybe only once in a lifetime, such as the decision to expand operations abroad.

Information might also be past, present or future. A decision to restock the supermarket shelf will be based primarily on past sales, but with a thought for the current buying pattern of customers. A decision to expand abroad would need a great deal of information about the future. Past information is much easier to obtain than information about the future, which is often based on opinion.

Activity 3

List five items of information which might be of no use if provided the day after they are required.

See Feedback section for answer to this activity.

Formal and informal information

A great deal of information we use is *informal*. It is stored in people's minds and its reliability depends on their level of recall whenever that information is required. The person receiving the information needs to be very careful as to how such information is interpreted. This kind of information is obtained through telephone conversations, over lunch and as chance remarks in conversation. It is often never recorded and so it is not available on a regular basis. Much of this information is based on opinion rather than fact.

Formal information, on the other hand, is recorded. It may be part of an accounting system, a personnel file, minutes of a meeting, etc.

It is difficult to say that formal information is necessarily better than informal information. The two often complement each other. After all, much of decision-making is based on the judgement of the person making the decision, even when faced with factual information, particularly at higher levels of management and especially when dealing with people.

List three types of both formal and informal information you receive in a typical day.

See Feedback section for answer to this activity.

The sources of information

The two main sources of information are *internal* and *external.* Internal information is gathered and provided from within the organization. It is on this kind of information that you base your day-to-day decisions, such as whether or not a doctor is free for an appointment at a particular time.

Internal information comes from systems such as your organization's accounting system, the appointments diary, the absentee records, etc.

External information comes from outside the organization. It is often used to make less frequent decisions, such as whether to increase the charge for a pint of beer. A decision such as this might need information on what price your competitors are charging - this is *external* information.

External information comes from sources such as government statistics, newspaper reports, and, increasingly, the Internet.

The communication of information

The most appropriate method of communicating information depends on a number of factors. These include:

- the cost of transmitting it
- the speed with which it is required
- the level of accuracy required
- the nature of the message, e.g. whether it is confidential
- the nature of the receiver, e.g. whether the receiver is internal or external to the organization
- the scale of the task, e.g. the number of recipients
- the importance of feedback from the receiver.

Information might be required by only one person, or by many. In addition, it might be required by a senior member of staff, or by a junior member.

You will learn more about different methods of communication in Chapters 6-9 of this book.

Quantitative and qualitative information

Quantitative information can be measured and expressed in numeric terms. Here are some examples:

1 Suppose the cost to the pub of a pint of beer is £1.00. If the pub down the road charges £1.50 a pint, you might decide to charge only £1.40 in order to attract customers. That is a profit of 40p a pint. Now suppose that staff wages per day are £40 – you can calculate that you need to sell 100 pints a day just to cover the wages, and pay the brewery for the beer.
2 If the doctor can see six patients an hour, and 60 people request appointments, you can calculate that the doctor needs to spend 10 hours in the surgery. That might mean 5 hours a day over 2 days. But if 60 people request appointments *every day*, it might mean the surgery needs a second doctor.
3 If the plane to Paris seats 200 passengers, and only 150 seats are booked, you can calculate that the percentage occupancy is only 75 per cent ($^{150}/_{200} \times 100$).
4 If the supermarket expects to sell 1000 tins of cat food a day, and the shelf holds 100 tins, then it needs to re-stock ten times a day.

There are a lot of numbers in the above examples. Numbers can be expressed in monetary terms (like the price of a pint of beer), in quantity terms (like the number of doctors needed), in percentage terms (like the occupancy of a plane), and in other terms such as fractions and decimals. Some of them might be better understood if they were presented differently, say in a diagram. You will look at ways of presenting information in Chapter 10.

Qualitative information is information which cannot be measured or expressed in numeric terms. It is often based on feelings or opinions. Examples include:

- customers who drink in the pub because it has a good atmosphere
- patients who are likely to arrive late for appointments
- passengers who use a particular airline because of the efficiency of the booking staff
- customers who shop elsewhere because the supermarket is always running out of cat food.

Both types of information are equally important to decision-making. It might be possible for the pub to increase its beer prices to £1.55 a pint and still retain the customers because of the lively atmosphere.

You can read more about quantitative information in *Resources Management*.

Look at the following statements. Which are examples of quantitative information, and which are examples of qualitative information?

1. 'The customers are always complaining about the slow service we give.'
2. 'Last month's sales exceeded target by 5 per cent.'
3. 'The food in our restaurant is of the highest quality.'
4. 'We expect a good turnout at next week's conference.'
5. 'I propose we give Janice a £500.00 pay rise.'
6. 'The current price of letterheads is £100.00 per thousand, plus VAT.'

See Feedback section for answer to this activity.

Characteristics of 'good' information

For information to be of optimum use, it must have certain qualities. Information which arrives too late, or is inaccurate, can result in poor decisions being made. It is important that the recipient is able to use the information, otherwise it is pointless providing it.

There are seven main characteristics which information should possess.

1. *Relevance*. The information should be of the right type for its purpose.
2. *Accuracy*. It should be true and reliable. It should also be to the right level of precision – the price of a pint of beer needs to be exact, but the number of tins of cat food on the shelf could be given to the nearest ten. In fact, large numbers are often better understood if they are 'rounded'. The population of a city of 12 836 472 would rarely be required to be quoted so precisely. If you were considering opening a sports centre there, your decision would not be affected by being told that there were 13 million people in the city.
3. *Timeliness*. Information should be communicated at the right time. If it arrives too late for a meeting, say, then an

incorrect decision might result. Equally, if it arrives too early it might also be useless – it could be lost or forgotten by the time it is needed.

4 *Currency.* Information should be current, i.e. up to date. An enquiry into the level of stock of a particular item should give today's figure, not last week's.
5 *Completeness.* All the required information should be provided, not just some of it. A little knowledge can be a dangerous thing!
6 *Clarity.* There should be sufficient detail provided, but not so much that it is difficult to extract the required items. If a manager requires a list of all customers owing more than £10 000.00, a list showing *all* customers who owed money would make the task of identifying the necessary ones much more difficult. Similarly, it might help if the customers were listed in order of the size of the debt, or the length of time the debt had been outstanding.
7 *Cost effectiveness.* If it takes a long time, or a lot of money, to provide information which is of little value, then it is not cost-effective. Surveying every pub in the locality over a 12-month period just to determine a price increase of 2 p a pint is unlikely to be cost-effective.

Investigate 1

Obtain three different pieces of information you give or receive at work. Check their characteristics against the 'good' characteristics given above. Identify any areas where the information could be improved.

Summary

This chapter has introduced you to the importance of information in organizations. It has identified the people who use information and the types of information which you, as a team leader, might need to use or to provide to others. You should now be better equipped to appreciate the qualities of good information and the need to ensure that it is communicated in the most appropriate manner.

Review and discussion questions

1 What does a team leader use information for?
2 Why is it important that information is properly managed?
3 List at least five ways in which information can be categorized.

4 What is the difference between formal and informal information?
5 List seven factors which affect the way in which information is communicated.
6 What is the difference between quantitative and qualitative information?
7 List eight characteristics of 'good' information.

Case study

Janet is a floor manager in a large, top-class hotel, with responsibility for the preparation of forty bedrooms on the second floor. Rooms are cleaned daily, and beds are changed on the third day, as well as before every new arrival. Soaps, towels and tea and coffee supplies are replaced daily. The hotel uses an outside laundry who normally return linen 48 hours after collection.

Janet has a team of four assistant housekeepers to help her, with three temporary staff to call on during periods of holiday or sickness.

In recent weeks, there have been a number of complaints via Reception of rooms not being ready for new arrivals, and twice the main storeroom has run out of supplies of soap. The laundry has now informed the hotel that they cannot return linen until 72 hours after collection. Today Janet overheard a guest commenting to his wife that the towels in their room had not been replaced.

Janet regards her team of housekeepers as reliable and efficient, and is concerned that she is not receiving the necessary information from others.

Identify the items of information which Janet needs in order to perform her job. Which items are internal and which are external? Can you suggest a more formal method of obtaining information on guests' opinions on the housekeeping standards?

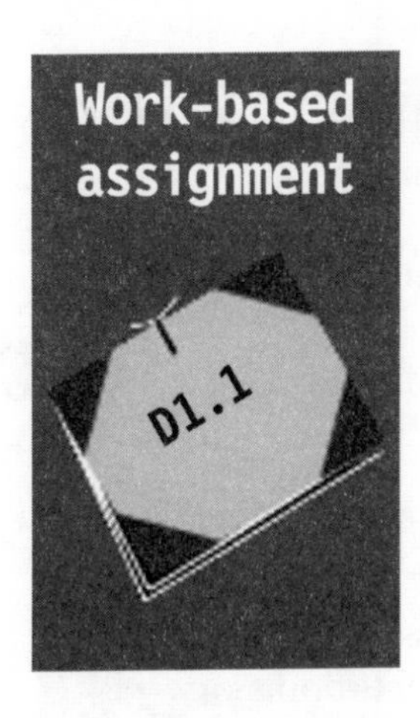

Select ten pieces of information that you receive or give during a day at work. Compile a table like the one suggested below and complete the columns as appropriate. The last column is for you to give each piece of information a score out of 7, according to how many of the characteristics of good information it contains. An example is given to help you.

Description of information	**Given or received?**	**Internal or external?**	**What used for?**	**Formal or informal?**	**Qualitative or quantitative?**	**Score**
List of contents of stationery cupboard	*Received*	*Internal*	*To compile order*	*Formal*	*Quantitative*	5*

* You might also note any qualities which were lacking, e.g. 'three boxes of envelopes' might be easier to grasp than '432 envelopes'.

2 Information for decision-making

Learning objectives

On completion of this chapter you will be able to:

- describe the levels of decision-making
- identify the types of information needed at each level
- describe the types of problem which occur at the different levels
- identify methods of solving these problems
- explain the effects of decisions at each level
- understand the importance of information for competitive advantage.

Introduction

Decisions are made at all levels within an organization, and at each level different types of information are required. Information also needs to be communicated to those who require it. The channels of communication are very much dependent on the type of information to be transmitted.

Organizations exist for many reasons, but in all organizations survival is important. Profit is not the only motive for organizations - indeed, many are not formed with the aim of making profit. But nevertheless there is competition between organizations for customers, market share, credibility and so on. Organizations who are able to make best use of information will achieve a greater recognition from their customers. This is known as *competitive advantage*.

This chapter will introduce you to the levels of decision-making and the information requirements at those levels, and the importance of organizations using information to improve their competitiveness.

Levels of decision-making

There are three levels at which decisions are made which correspond to the levels of management within a typical organization. They are:

1 strategic level (top management)
2 tactical level (middle management)
3 operational level (junior management).

These levels can be portrayed by the *Anthony Triangle*, shown in Figure 2.1. The shape implies that there are fewer decision-makers and decisions made at strategic level than at operational level.

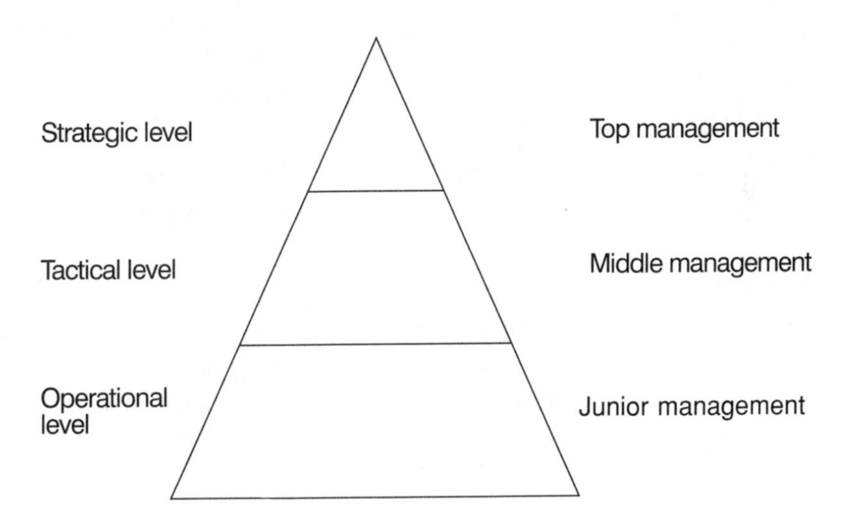

Figure 2.1

As a team leader, you are likely only to be involved in decision-making at operational level, but it is useful to understand the different types of decision made at all levels, as they may have an impact on the decisions which you are asked to be involved in, and your decisions may also affect those at a higher level. Of course, there is often overlap between the different levels.

The decisions you are involved in will also vary depending on the structure of your organization and the grouping of work teams. If you are involved in a special project, for example, you may find yourself working with managers at different levels, and will need to understand the information needs which they have.

In this chapter, you will concentrate on looking at operational decisions. As a contrast, the chapter also includes some examples of strategic decisions. Tactical decisions usually require a combination of both operational and strategic decisions.

Information requirements at the different levels

Information requirements are portrayed in Table 2.1.

Table 2.1 Information requirements at different levels

Characteristic of information	Strategic level	Tactical level	Operational level
Source	External	External and internal	Internal
Scope	Very wide, not defined	Intermediate	Narrow and well-defined
Level of detail	Broad, summarised	Partly summarised	Detailed
Currency	Future	Future, present, past	Past
Timeliness	Not urgent	Some urgency	Immediate
Frequency	Infrequent (annual)	Regular (monthly)	Frequent (daily/hourly)
Level of precision	Low, approximate, qualitative	Fairly precise, quantitative and qualitative	Exact, quantitative

Assume you are a manager in a sports and leisure centre. Draw up a table showing each of the following decisions, and identify their requirements, guided by Table 2.1:

1 A decision to build a 'sub-tropical' swimming pool with slides, wave machines, water sprays etc. and pool-side refreshment facilities.
2 A decision to allow local schools and youth groups discounted admission fees.
3 A decision to reorganize the lifeguard staff rotas.

See Feedback section for answer to this activity.

Information requirements at operational level

Let us look in more detail at the information required at operational level.

Internal information

Internal information comes from sources such as reports, machine readings, stock listings, time sheets and appointments diaries. Often the information is produced in the department in which you are working, by the staff whom you supervise. It follows that you have a high degree of control over this information, and its reliability should be known to you. You should be able to access it quickly and easily.

Imagine you are the supervisor in the parts department of a large car dealer. Your information will consist of stock records, supplier names and addresses, customer orders received and the location of the stock in the storeroom. You will also have a product-description file. You might also need a customer listing for any customers who buy on credit (i.e. they do not pay at the time of buying).

Narrow and well-defined information

You will know exactly what information is required for the decision you are making. If a customer requires a part for his car, you will know to ask the make and model, the year of manufacture, a description of the part, when it is required, whether it is in stock and where to order it from if it is not.

Detailed information

You will need a high level of detail to make your decision. The stock records should list every part available, the product descriptions should be full, customer orders should give full information and customer listings should contain all relevant customers.

Past information

This is also referred to as *historical* information, because it is based on activities which have already taken place - even though the decision is being made now. Thus, stock records will contain details of stock transactions which have already occurred, orders that have already been received, customer listings will include previous customers and the amount of credit they can have. If you are responsible for budgets and targets, you will also be comparing your current sales with previous sales.

Immediate information

The information you require must be available immediately. The stock listing will need to be at hand to answer a customer telephone enquiry. The customer listing will have to be referred to before allowing a customer to take goods without paying. The price list will also have to be accessible at once.

Frequent use of information

You will need this information daily or even more frequently. Customer enquiries will be coming in all the time, many of them without prior notification, and they will expect to be answered at once.

Exact information

Information must be precise at this level, often expressed in numeric terms, i.e. quantitative information. It is no use knowing that there are 'about ten' items of a particular part in stock, if a customer requests 11.

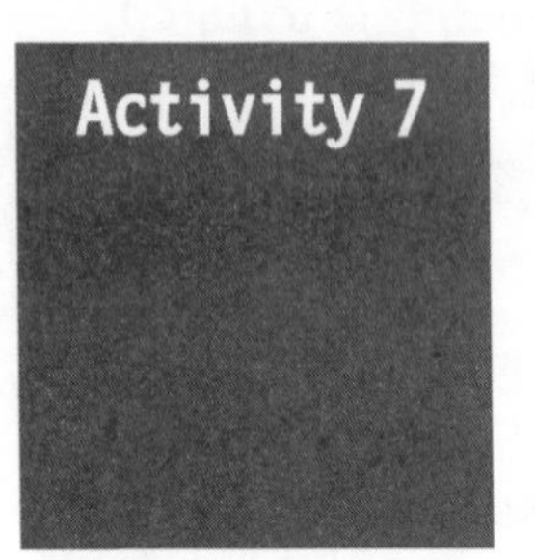

Imagine you are gathering together the information to determine whether or not to employ additional staff over the Easter holidays to cover the leisure centre mentioned in Activity 6. Using each of the categories of information above, identify the possible information you would need to enable you to reach such a decision.

See Feedback section for answer to this activity.

Making decisions with the information available

Table 2.2 helps to portray the different types of decision.

Table 2.2 Making decisions with the information available

	Strategic level	Tactical level	Operational level
Type of problem	Complex	Semi-complex	Simple
Method of solution	Unstructured	Semi-structured	Structured
Effect of decision	Long-term	Medium-term	Short-term

Type of problem

At operational level, the types of problem are relatively simple - although there are prpbably many team leaders who have found this not to be the case at times!

A typical problem in the car-parts department might be a customer request for an item which is out of stock. That is a simple problem, but of course it can become more complex if the usual supplier is unable to supply, or if the part has been modified and will not fit the car. In such cases you might need to refer upwards to your manager for help.

At strategic level, problems are much more complex. In fact, in many cases, management do not even know what the problems are! Take the example of expanding operations into Europe. As mentioned in the previous chapter, in contemplating such a decision management will need information on world markets, foreign currencies, international trade laws, etc. They might not realise until they are part-way through their investigations that the company's product has to be modified to suit the foreign market. They might find that their British employees object to the proposal because of fear of losing their jobs. They might discover problems with transporting supplies to the new location. The shareholders may veto (vote against) the move. Many of these problems do not surface immediately, and are much more difficult to solve than that of a parts shortage.

Method of solution

A *structured* solution is one where it is possible to follow a set pattern of 'steps' in order to solve the problem. For the above example of a customer requesting a part, the steps might be:

- check the stock records for availability
- offer the customer the items in stock
- consult the supplier listing
- enquire of the supplier the timescale to obtain new stock
- pass this information on to the customer
- order new stock
- await delivery of the stock

- allocate the stock on arrival (to make sure it is not sold to another customer)
- inform the customer of its arrival.

Most operational decisions can be taken using a structured approach. They have often been made many times before, and there may even be a Procedures Manual which tells staff what to do.

At strategic level, the solution is *unstructured*. Although there may be some structure to the individual parts of the problem, there is unlikely to be any set procedure for the whole. It is also likely that management have never faced this problem before and are unsure as to how to go about finding the solution. Many of the solutions are based on qualitative information, and depend on the judgement of management rather than on hard facts.

Investigate 2

Take a structured decision made by you during your working week and detail the procedures you would go through to reach a reasonable conclusion.

Effects of decisions

The effects of operational decisions are usually short-term. Fulfilling a customer's order is forgotten once it has been done and you move on to the next. If you make a mistake, say in giving a customer 12 items and charging only for 11, the effect is (hopefully!) minimal. Obviously, you must not make too many of these mistakes, but you are not likely to ruin the entire company by such an error.

At strategic level, however, decisions are likely to affect the company for a very long time, and mistakes will be highly expensive to correct. It may not even be possible to correct them. You can read more about short-term and long-term decisions in *Resources Management*.

Information for competitive advantage

Every organization wants to survive. A lot of people, and other organizations, depend on it for their survival too. That includes its customers, suppliers, employees, owners – in fact all of those 'users of information' you looked at in the previous chapter.

Organizations cannot stand still. In today's environment they must look for new methods and procedures, new markets, new products and services. In some cases, they must first look to retaining what they already have, as others fight to survive themselves. In short, organizations are all rivals with one other for survival. 'Competitive advantage' means using information to ensure that your organization stays afloat, and improves its position in the marketplace.

Improving the chances of survival is not just a senior management task. It falls to all levels of staff to play their part in ensuring the company's survival. If you deal with any of the areas shown in Figure 2.1, then you are able to influence the survival chances. For example, if you have telephone contact with customers you can make sure that you maintain a good relationship with them and that any information which they pass to you about your products and services, and the products and services of others, is recorded and acted upon.

Information helps organizations to keep afloat in the market place. Some examples of where information can be useful are:

- identifying new competitors
- identifying new products
- determining the needs of customers
- identifying the strengths and weaknesses of suppliers
- keeping abreast of new materials, methods and technology
- ensuring that you have the most accurate and up-to-date information
- using information technology to aid decision-making.

Investigate 3

Identify and describe a problem, the solution to which could improve competitive advantage in your organization, from something for which you are responsible. What information do you need to respond to this problem, and how can it contribute to your competitiveness?

Summary

This chapter has introduced you to different types of decision made at the three different management levels. In particular, it has looked at decisions made at the operational level, which is the area in which you will most likely be working as a team leader. The chapter has also considered the type of approach to solving methods at this level, and the effects of decisions made. You have also been introduced to the concept of information as important for competitive advantage, and have seen how all levels of management can contribute to this.

Review and discussion questions

1 Describe the levels of decision-making.
2 Identify the types of information needed at operational level.
3 Describe the types of problem which occur at the different levels.
4 What method of solution to problems is most appropriate at operational level?
5 What effects do decisions at operational level have on an organization?
6 What is meant by 'competitive advantage'?
7 What areas of your job enable you to contribute towards 'competitive advantage' for your organization?

Case study

Mahmood is a team leader managing four bread-makers in a bakery. During the course of a normal week, he has the following duties to perform:

- obtaining the ingredients for each bread-maker from central store
- ensuring that the ovens are in good working order
- supervising the cleaning of ovens between each batch of bread
- assessing the quality and appearance of bread produced
- recording the quantities produced, with a target provided daily
- arranging delivery of the finished bread to the warehouse.

Answer the following questions regarding Mahmood's job:

1 At what level of management is Mahmood working?
2 What are the characteristics of the information which Mahmood is giving and receiving?
3 How might Mahmood solve the problem of the flour supplier notifying him that he is unable to deliver the usual quantity of flour next week? Consider the *structure* of the solution.
4 In what ways might Mahmood be able to contribute to the competitive advantage of the bakery?

Choose a decision in your workplace which you have had to make before. Describe the decision. Identify the items of information you need to make the decision and list their characteristics:

- what are their sources?
- how narrow and well-defined are the items of information?
- what level of detail is required?
- is the information past, present or future?
- when is the information required?
- how often do you need this information?
- how precise does it need to be?

Describe how you are going to make the decision, including the steps you need to take. Do you need to refer to any manuals or written procedures? Does your decision have a long-term or short-term effect on the organization?

During the task, did you have any opportunity to improve relationships with customers, suppliers or staff, or to find out about new products or methods which might help your organization in the future?

3 Gathering data

Learning objectives

On completion of this chapter you will be able to:

- explain the difference between *data* and *information*
- understand the basic *data processing model*
- choose and use a variety of different methods for gathering data
- assess the reliability of data gathered
- explain the purpose and basic principles of the Data Protection Act 1984.

Introduction

It is important that the information used by people is reliable. This means that great care needs to be taken in gathering it, otherwise decisions may be made which are incorrect. The Americans have an abbreviation which they use - GIGO - which stands for 'Garbage In, Garbage Out' - in other words, if you put rubbish into something then the end result will be rubbish. Organizations cannot afford for their decisions to be rubbish, so it is vital that what goes into their information system is of high quality.

This chapter tells you about the distinction between *data* and *information*, and introduces you to methods of gathering data which can be used to provide useful information.

Data and information

In everyday language you probably use these two words to mean the same thing. But there is a distinct difference between them.

Data is the starting point in providing information. It is the facts, figures, values or even opinions which are needed to aid decision-making. It is often referred to as 'raw' data, which implies that something needs to be done to it before it is useful - rather like cooking a joint of meat before you can eat it.

This raw data needs to be *processed* in some way before it becomes *information.* You will be told in Chapter 5 that information is knowledge which is meaningful and useful to its users. Data is not always so.

The basic data processing model is shown in Figure 3.1.

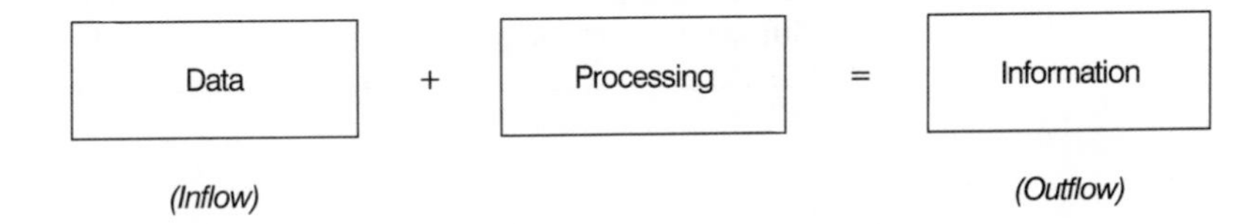

Figure 3.1

Look at an example. Suppose you want to find out the ages of students taking different courses in a college, to find out if particular courses attract different age groups. You could stand by the college door one day and, as each person comes in, ask their age and the course they are studying. You would write all these down on paper, as they arrive. At the end of the day you would have several pages of data in no particular order. Your list might look something like this:

Age	**Course**
21	Childcare
25	Business studies
42	Hairdressing
19	Business studies
17	Hairdressing
26	Supervisory management
28	Childcare
24	Supervisory management
31	Business studies
27	Supervisory management
53	Hairdressing

and so on.

This is a list of *data.* In order for it to be meaningful, it needs to be summarised. You might do this by producing a table, as in Table 3.1.

Now you have information! You can see that Supervisory management attracts mainly older students, whereas Childcare attracts mainly those under 20. Hairdressing attracts those under 30, whilst Business studies has a greater spread of ages, but mainly younger students.

Table 3.1 Number of students in age range

Course	Under 20	20–29	30–39	40 or over
Business studies	25	15	8	4
Childcare	21	7	2	nil
Hairdressing	18	15	nil	nil
Supervisory management	2	10	22	16

You could use this information to decide how to market the college's courses to these groups, or which classes to circulate regarding a coach trip to see a famous rock band.

The raw data has been sorted, classified (into age groups), and summarized by both course and age group. This is *processing*. You might have also totalled the rows and columns - performing calculations is another example of processing.

The information has then been *presented* in a particular manner (i.e. in a table), to enable it to be *communicated* to the person who needs it.

A final stage is to *analyse and interpret* the information, and to decide whether you need more, or different, information in order to make your decision.

You will look more closely at presentation and interpretation of information in Chapter 6.

The whole model can be shown as in Figure 3.2.

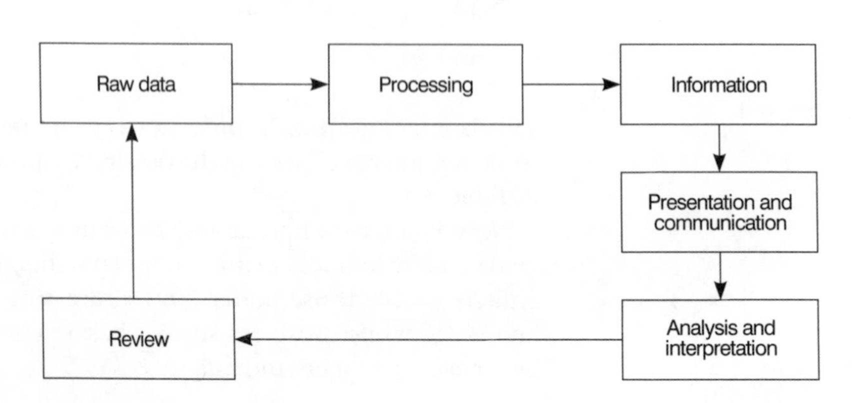

Figure 3.2

Which of the following are examples of *data*, and which are examples of *information*?

1 A list of all items in stock, in stock number order.
2 A list of stock which has fallen to a low level.
3 The names of all patients with hospital appointments during next week.
4 A completed appointments diary.
5 A total of expenditure in your department, broken down into sub-totals for stationery, postage, telephone costs and equipment hire.
6 A file of invoices and receipts for the expenditure in 5.

See Feedback section for answer to this activity.

Locating data

The first problem in collecting data is to find out where it is! It is a sad fact that much data collected by individuals has already been collected by someone else - and therefore a lot of time is wasted in repeating the exercise unnecessarily. A second sad fact is that a lot of data which is collected is never recorded, and so the exercise *has* to be repeated.

The three main locations of data are:

1 Data already recorded in the organization. This may be held in manual files, magnetically (e.g. on tapes or microfilm), or electronically (on computer). It may still need processing for any particular purpose, but at least it has already been gathered.

   It is important, though, to make sure that the data is accurate and up-to-date before relying on it. It may have been gathered some time ago and may need updating, or it may have been gathered for a specific purpose, in which case it might not be quite what is needed. Remember the term 'GIGO' before you use it.

   Data which is gathered for a particular purpose is called *primary* data. Data which has been gathered for some other purpose is called *secondary* data. Generally, primary data is more reliable than secondary data.

2 Data available within the organization, but not recorded. This includes data gathered orally at meetings or during conversations, data gathered through observing people and processes, and data held in people's minds such as knowledge, opinions and experience.

3 Data potentially available from outside the organization. This may be recorded or unrecorded. If it is recorded, you need to be able to obtain those records; if it is not recorded, then you will need to gather it yourself. The main difficulty with obtaining data from outside the organization is its cost. You may have to pay to gain access to recorded data, or take up a great deal of time in gathering it for yourself. Remember, one of the qualities of good information is that it does not cost more to provide than its value to the organization.

Investigate 4

What data do you use in your job which already exists within your organization? Which data is already recorded and which do you have to collect yourself? What data *might* you be able to make use of from outside the organization?

Data from information

Information which has already been processed might become data for another purpose. For example, a customer's account will already have been processed with details of goods bought, returns, payments made, refunds given etc., and will show the present amount outstanding and the length of time the debt has been owing. This could provide data for the credit control supervisor in chasing up outstanding debts. The data might need to be classified first into accounts outstanding for less than 30 days, those outstanding for between 30 and 60 days, those between 60 and 90 days, and those over 3 months old. The action to be taken will depend on the length of debt. Those under 30 days will probably be left for the time being; those between 30 and 60 days will get a pleasant reminder letter; those betweeen 60 and 90 days will get a stronger letter, and those over 3 months old will be forwarded to the solicitor for action.

Data collection methods

There are many different methods of collecting data. The method you choose will depend on these factors:

- how much data is needed
- where it is coming from
- how accurate it needs to be
- how quickly you need it
- the cost of obtaining it.

Interviewing

Interviewing is a relatively slow and time-consuming method of obtaining data. You can only interview one person at a time, and you may have to travel some distance to each person. Because of the time and cost involved, you should make sure that your interview is well planned beforehand so that you get maximum benefit from it. You also need to choose carefully who you are going to interview. If you are only able to choose a few people, you should select those who are in key positions.

On the other hand, interviewing has the advantage that you can interact with the interviewee and they with you, so you are able to ask additional questions or give clarification on any questions which are a little vague. You can make sure that there is no misunderstanding on either side. Interviews are a good way of obtaining opinions or further data on areas that you are unsure about.

It is important that the interviewee knows the purpose of the interview, so that they can give the clearest and most relevant answers to your questions.

Chapter 7 gives you more information about interviewing techniques.

Observing

This is another slow and time-consuming method, but is often the only way of finding out *how* people do things. However, do bear in mind that people often behave differently when they are being watched. They might normally take 'short cuts' when no-one is around, or perform at a different speed. They might be nervous about being observed.

As with interviews, you should explain the purpose of your observation to the person - it is not fair to observe people secretly - even though you might think you will get a more accurate view of their activities! Later in this chapter, you will read about the Data Protection Act, which states that data should be obtained *fairly*. Sneaking up on people is not fair!

Questionnaires

These are used where data is required from a large number of people. It is possible to issue them far afield and you do not need to be there when the person completes the questionnaire.

They are, therefore, more cost-effective than interviews and observations. However, unless you are able to encourage the person in some way, the rate of response to questionnaires is quite low – many end up in the recipient's bin.

Questionnaire design is quite difficult. You need to make sure that it is not too long and is easy to complete. Those with boxes to tick are easiest, or those which require only a Yes or No answer. However, it is important to give the person sufficient choice in the answers. For example, a question such as 'Do you eat chocolate?' might be given a Yes/No option. A person who eats it sometimes will find it difficult to choose which box to tick. Giving four options is often useful, e.g. every day, once or twice a week, occasionally, never.

It is also important to make sure that the questions are unambiguous, i.e. their meaning is clear. A question such as 'How often do you take holidays?' might be interpreted as how often do you *go away* on holiday or how often do you take time off work for any reason.

Open-ended questions can be used to allow people to add relevant comments, or to give them complete freedom in their answers, but they are difficult both to set and to analyse.

Completing forms

Much factual and numeric data can be gathered by the use of forms. You will have completed forms many times both in your job and in your everyday life. There are forms for ordering goods, applying for bank loans, job applications, tax returns, etc. In the workplace there are forms for claiming overtime pay, recording staff sickness, ordering stationery, listing invoices, etc.

Good form design is essential. Forms should be easy to complete and follow a logical sequence of completion. You should not have to keep turning pages forwards and backwards. Some of the features of a good form are:

- sufficient space to complete fully
- clearly labelled boxes and sections
- no repetition of the same data in different places
- paper of sufficient quality, both to write on and to survive whatever is going to happen to it after completion
- duplicate copies where required
- pre-printed data wherever possible
- clear instructions for completion.

Bear in mind the purpose of the form. If it is to be used for input to a computer, numeric data might need to be in a fixed format, e.g. pounds and pence, or the date written as 24/01/98 rather than 24 January 1998.

Chapter 4 gives more information about how data is input after it has been collected.

Activity 9

Design a form to record a mail-order customer's order. Consider the features of good form design above. How many copies would you need? Give instructions for the completion of the form if you think this is needed.

See Feedback section for answer to this activity.

Investigate 5

Obtain samples of three forms from your workplace. Critically assess each one as to its features. Complete a table like the one below:

	Form 1	Form 2	Form 3
Title of form			
Brief description			
Sufficient space?			
Clear labelling?			
Repetition?			
Paper quality?			
Duplicate copies?			
Pre-printed data?			
Clear instructions?			

Activity 10

What method of gathering data would be most appropriate for the following:

1 ascertaining the opinions of staff on the new canteen facilities
2 determining the hygiene procedures in a kitchen
3 recording the details of a new customer in a bank
4 investigating a complaint about a member of staff.

See Feedback section for answer to this activity.

The reliability of data gathered

Remember GIGO again! Not all data gathered is totally reliable.You have already learnt how people change their behaviour when being observed. They do not always answer questions accurately, either deliberately or mistakenly, or the questions may have been unclear. Even factual information is open to error when it is recorded. The fact is that people make mistakes and you need to bear that in mind when gathering data. Data which needs to be error-free will need to be checked before it is processed, and probably afterwards too.

Recording data

Data gathered manually must be recorded. Always keep records of interviews and observations, perhaps on tape or camera, but certainly you should record your findings as soon as possible after the event or, better still, as the interview or observation proceeds. Otherwise you are in danger of forgetting what you saw or heard.

Data gathered electronically or automatically is more accurate. The use of bar-code scanners in supermarkets and elsewhere reduces the risk of error. Forms with pre-printed data are also more reliable. In Chapter 4 you will learn about electronic methods of inputting data.

Acceptability of errors

Some degree of error might be acceptable, provided that it does not affect the decision which results from the data. For example, counting the quantity of stock as 432 instead of 433 will not materially affect any orders you accept, or the picture given by the financial accounts (unless, of course, it is stocks of gold bars!). It is a popular misconception that accounts which are audited are totally accurate, in fact an audit only confirms that the accounts give a 'true and fair view' of the organization's affairs, and minor errors are not considered to affect this.

The cost of being totally accurate can be higher than the value of the information produced.

Sample sizes

It is not always possible to gather all the data available, and not always necessary either. Sometimes you need to consider taking only a sample of the possible data and basing your decisions on that. Your organization might have 20 000 customers. If you want to obtain their views on after-sales service it would be impossible to survey all of them. Issuing a questionnaire to them all would probably result in only a small proportion being returned - and it would tend to be those who had complaints. The customers who were happy with your service, or who had not needed it, would not be inclined to respond. It might be better to choose a smaller sample from all three types of customer and perhaps offer some incentive for completion, such as a discount voucher. Such a sample of, say, 1000 customers would more than likely give you a reliable view of all customers' opinions.

The Data Protection Act 1984

This piece of legislation was introduced to protect individuals who had data about them stored on computer systems. It was feared that, because computerized data can be transmitted easily and might be accessed by people other than those who had collected it, individuals were entitled to some assurance that the data would not be passed on without their permission.

It is likely that this legislation will be extended to cover data stored on manual systems in the near future.

Data subjects

The Act only applies to data about individuals, not about companies or organizations. In addition, it only applies to living individuals, and they must be capable of being identified by the data which is stored about them. Such people are known as *data subjects*.

Data users

Data users are those who store, process or just control the data. So it covers people who do not have computers themselves but who use computer bureaux or other organizations to perform the work for them. It also covers not only computerized data, but other types of automatic equipment. But at present, it does not cover wholly manual records.

The data user must register with the Data Protection Registrar.

Activity 11

Explain what is meant by the terms *data subjects* and *data users*.

See Feedback section for answer to this activity.

Exemptions

Some types of data held for particular purposes are exempt from the Act and need not be registered. These include:

- data held for household purposes, such as the names and addresses of your friends and family, or data needed for household accounting
- payroll, pension and accounting data, used to perform calculations of pay or pension, or to keep accounting and general business records
- membership lists for clubs and societies involving names and addresses and subscriptions details. However, members have to be asked if they object to their data being held electronically, or if they object to their data being passed to anyone else.
- data held for statistical or research purposes
- data which is required to protect national security.

Which of the following would require an organization to register under the Data Protection Act?

1 A college holding previous qualifications of its students.
2 A hospital holding data about a patient's next of kin.
3 A builder holding customers' names, addresses and details of work carried out.
4 A business holding data about employees' training records.

See Feedback section for answer to this activity.

The data protection principles

Any data user who registers must agree to the following eight principles, that all personal data must:

1 be collected and processed fairly and lawfully
2 be held only for specific and lawful purposes
3 be used only for those purposes and only disclosed to persons as stated by them when they register
4 be adequate, relevant and not excessive for the purposes
5 be accurate and, where necessary, kept up to date
6 not be kept longer than is necessary for the purposes
7 be disclosed on request to the data subject at reasonable intervals, without undue delay or expense, with the individual being entitled, where appropriate, to have such data corrected or erased
8 be kept secure against unauthorised access, alteration, destruction, disclosure or accidental loss.

On registration, the data user must state what data it is intended to hold, why it is being held, and who, if anyone, it is to be disclosed to.

The Data Protection Registrar can refuse to register an application, amend it, or withdraw it if any of the principles are not adhered to.

As a supervisor, you may well be responsible for collecting, storing and using such data, and you need to know whether or not your registration is affected by your activities.

Is your organization registered under the Data Protection Act? If not, can you identify any data which *ought* to be registered (because it is held on computer and is not exempt) or which would *need* to be registered if it were held on computer?

Look at any records which you are responsible for maintaining. Are any of them registered? If so, what does your registration allow you to do with the data (you will need to check your 'entry' in the Register to determine this).

If you feel your organization is in breach of the Data Protection Act, notify your manager at once!

Summary

This chapter has looked at the difference between *data* and *information*, and has explained that data needs to be processed in some way in order to become meaningful information to its users. This is portrayed by the *data processing model*. The chapter has also looked at different methods for gathering data, and the reliability of data being gathered, including the use of sampling. The provisions of the *Data Protection Act 1984* have also been outlined to enable you to determine the need for your organization to register any of the data which it holds.

Review and discussion questions

1 What is the difference between *data* and *information*?
2 Draw the basic 'data processing model'.
3 What is meant by 'processing'? Give examples of ways of processing.
4 What stages follow after processing?
5 What are the five factors to be considered when choosing a method of data collection?
6 What are the four common methods of data collection?
7 Is it always necessary that data should have no errors in it?
8 What are the eight data protection principles?

Case study

Lesley is a team leader in a travel agency. The agency deals mainly with two types of customer – those who like to book cheap, reduced-rate holidays at short notice, and those who book ahead. Of those who book ahead, some like to take brochures home and decide on their holiday by themselves; others need a great deal of guidance in choosing their holiday. Every week she needs to check the availability of holidays, both for the coming week and for months ahead.

Identify the data that Lesley might need in order to help her fulfil customers' requests. Where would she get these data from? How could they be converted into information which she would find useful? What methods would she use to determine customer requirements? Can you think of any circumstances which would require her to register under the Data Protection Act?

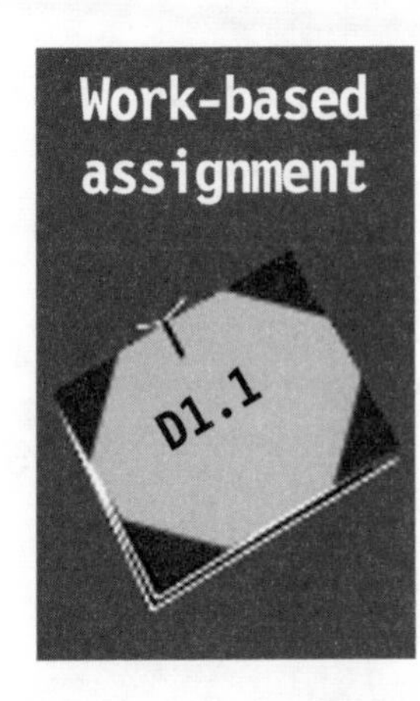

Choose a system or procedure in your organization with which you are involved. It need not be a computerized system. Write a brief description of the system, to include the following points:

- an outline of the system/procedure – what it is intended to produce?
- where does the data for the system come from?
- how is the data gathered? Could you improve on the methods used?
- how reliable are the data?
- does it comply with the Data Protection Act?

4 Inputting and processing data to produce information

Learning objectives

On completion of this chapter you will be able to:

- describe manual systems for recording and storing data
- describe computerized systems for inputting and processing data to produce information
- describe the hardware and software components of computerized systems
- explain the advantages of computerized systems
- identify hardware for particular tasks
- understand the purposes of different types of software
- explain methods of input, processing and output in computer systems.

Introduction

This chapter explains the basic methods of recording and storing data, both manual and computerized. It discusses the advantages and disadvantages of the different methods. Despite the increased popularity of computerized systems, manual systems are still heavily used in both large and small organizations, and their importance should not be ignored. Many of the principles of recording and storing data apply equally to manual and computerized systems.

The hardware and software components of computerized systems are undergoing very rapid development, even as this book is being written.

Manual systems for recording and storing data

Many organizations still rely heavily on manual systems, including those which have computerized systems in some areas. The first computers were used for the control of machinery and it was much later that they were used for providing information. Manual systems include the following:

- card indexes
- directories
- catalogues
- ledgers
- files of documents.

The word 'manual' means 'by hand', but in this context it also covers typed and printed records.

Financial data

Financial transactions occur daily and can be recorded in manual lists maintained by the accounts department. Examples include purchase invoices, sales invoices, credit notes and the receipt of cash or cheques in payment of debts. Purchase orders may be dealt with by the Purchasing department. Stock records are updated by the storekeeper. Ledger accounts of customer and supplier transactions are maintained by the accounts department.

In many small businesses these records are maintained manually and, for small businesess with few transactions, this is perfectly acceptable. The author of this text curently maintains the accounts of a sole-practitioner solicitor on a manual basis and this is the most cost effective means available.

However, manual systems have their limitations. In the case of the sole practitioner solicitor, if she wants to know who owes her payment for drawing up a will, the book-keeper has to get out the Wills record and manually list the wills prepared for which payment has not yet been received. This can be a very substantial and time-consuming task. It is also open to mistakes; manual records are error-prone. For example, the book-keeper might miss a bill from the list.

Other data

Organizations keep manual records of other items such as customer names and addresses, catalogues of goods for sale, employee records, the location of machinery – the list is endless.

Many of these records are duplicated. The customer list will be needed by the accounts department to prepare invoices and to send out reminder letters. It will be needed by the despatch department to send out goods. Sales representatives will need it to contact their own customers. The marketing department will need it to carry out mail-shots and for special promotions. If there is a central manual file, then staff have to leave their desks to go to it - and it could be a long way away. In addition, there could be a dispute over the order in which the file is kept: the sales representatives want it filed according to sales area, marketing want it filed according to the type of customer or the products that they buy, the accounts department want it filed alphabetically.

So, in many cases, each department keeps their own copy. When changes are made, such as adding a new customer, or amending an address, every copy needs updating. This is time-consuming and again, mistakes can be made.

Manual records also take up a great deal of space.

You can read more about methods of recording in *Resources Management*.

Activity 13

List the disadvantages of manual systems for recording and storing data.

See Feedback section for answer to this activity.

Investigate 7

Identify four systems in your organization which are maintained manually. Give a brief description of each and identify any problems which you can see exist.

Electronic systems for inputting and processing data to produce information

Whilst there are many useful types of machinery in use, such as tape-recorders, typewriters, photocopiers, telex machines, fax machines, microfilm equipment, calculators, etc., some of them with storage or memory facilities, this section deals primarily with computerized systems.

Types of computer

In the 1950s and 1960s, the use of computers in large companies grew steadily. These computers were *mainframe* computers. They were very expensive, often took up a whole room to accommodate, and a company would have only one, centrally located, with a department of specialist staff to operate it. Data was collected and organized by the staff in the user departments and input by a team of keyboard operators. Then it was transferred to large tapes or discs which were fed into the computer to update the records. The output would often be large amounts of computer printout on paper, which went back to the user department.

Mainframes are still in use but have developed tremendously from those of the 1960s, which were less powerful than some of today's electronic calculators! During the 1970s a slightly smaller computer, known as the mini-computer, was developed. During the 1980s there was rapid development in much smaller computers, with increased power and capabilities. Most organizations now use these desk-top microcomputers instead of, or as well as, a mainframe computer. Some are known as 'personal computers' or 'PCs' for short. In addition, the use of portable computers, laptops and even smaller 'notebook' computers is on the increase. All of these perform in the same basic manner as the mainframe computer.

The component parts of a computer system

You do not need to know how individual parts of a computer work, but you should have an understanding of the purpose and use of the different components. The computer can be depicted as in Figure 4.1.

Apart from the data being input and the information being output, the two main components of a computer system are *hardware* and *software*.

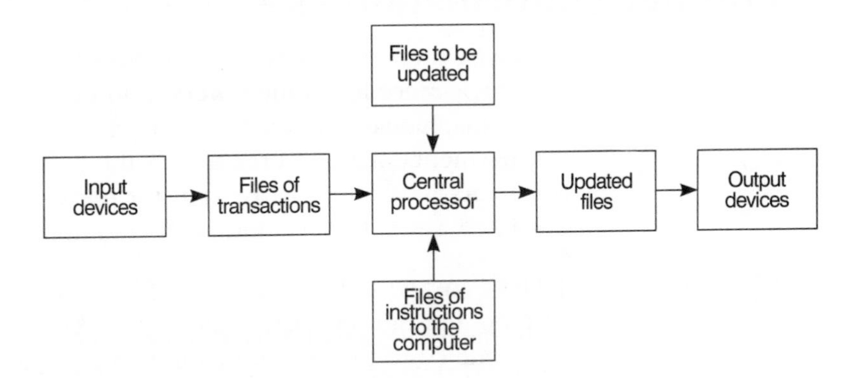

Figure 4.1

Hardware is the tangible equipment which is used. This includes the input and output devices, the tapes and discs of data and information, and the central processor itself, which converts the data into information. But the computer does not do this by itself. It needs to be told what to do. This is the *software*. Software is the program of instructions by which the computer operates.

The system also uses a number of *files* during processing.

Files

A computer system uses a variety of files in the processing of data.

Master files

These are the files that hold the finished result. A master file might be a file of stock records, holding all the data about stock bought, sold, returned, scrapped, etc. for each item of stock. Other master files might be:

- the sales ledger (debtors file)
- the purchase ledger (creditors file)
- the payroll file
- the fixed asset register.

The data in these files change regularly, and the files therefore need updating with the changes.

Transaction files

These are files of changes to be made to other files. Examples would be:

- stock movements for a week
- invoices issued to customers during the day
- payments made to creditors at the end of the month
- hours worked by employees for the week
- fixed assets bought during the month.

The transaction file and master file are both fed into the computer and the master file is updated either to replace the old one or to produce an entirely new master file.

Reference files

These are files of 'fixed' or standing data which do not change frequently, but which might need to be referred to while updating the master files. Examples include:

- descriptions of stock items
- customer credit limits
- creditors' names and addresses
- hourly rates of pay and tax codes
- depreciation rates.

Program files

These are the instructions for the computer to perform the required tasks. Writing programs is highly specialized and complex. In Chapter 5 you will look at some areas where users can write their own programs, but most programming is carried out by experts.

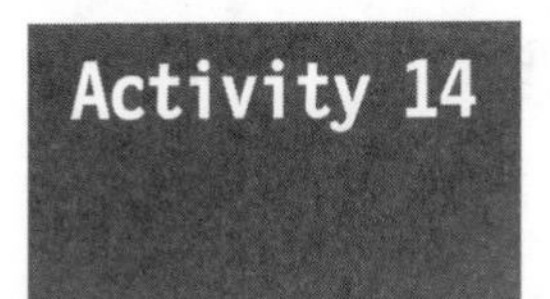

What are the four types of file found in a computer system? In a system which is used to calculate wages, what files would be needed?

See Feedback section for answer to this activity.

Hardware – input devices

These are needed in order to get the data into the computer.

The keyboard

This is probably the most common method of entering data. The keyboard operator might have a pile of invoices, or a customer request for an item of stock. The keyboard is a relatively slow method of input and is prone to errors.

The screen or VDU (Visual Display Unit)

The screen can be used as an input device itself, using a pointer or even a finger to touch the screen.

If the computer uses Windows systems (i.e. it has pictures and symbols on the screen), a 'mouse' can be used to point to the symbols and to input options. The mouse is a hand-held device which moves an arrow-head around the screen.

The screen is also commonly used to check the input from the keyboard and errors can be amended as they are spotted.

Bar code readers

These are common in shops. The bar code is a series of black stripes affixed to a product, representing a code. When scanned into the reader, the computer recognizes the code and is able to use it to bring up the price and a description of the product. But its use also has other advantages. There is no need for the price to be affixed to each item, as it is stored in the computer. If there is a price change, the computer is simply amended - there is no need to re-price all the items on the shelf. It can also be used to automatically update the stock records.

A major advantage is that, because there is no human intervention, the risk of error is almost nil. The scanning process is also very fast.

Bar codes are also used in other areas, such as on employee identification cards.

Character readers

There are several of these around. Perhaps the most familiar are the preprinted numbers on your cheques, giving the bank code and your account number along the bottom. When the cheque is passed through the bank, the value of the cheque is encoded in the same way, and read into the computer automatically. Some organizations use similar equipment to produce customer invoices; the computer produces the document with certain items of data pre-printed; the customer receives the bill, returns it with their cheque, and it is read back into the computer with the details already included on the document. This is known as a 'turnround' document.

Obvious advantages are speed and a lower error rate.

Mark sensors

These are used by filling in boxes with solid blocks or other marks which can be then read into the computer automatically. Common examples are questionnaires and multiple-choice exam questions, where the student fills in one of four boxes with a pencil. Again, speed and a lower error rate are advantages.

Document scanners

These are able to scan in a whole page at once. They can cope best with plain text, but latest models can scan graphics and photographs. The time saved is obviously great.

Magnetic stripe cards

These are commonly used on credit cards and cash cards. Your personal number is held within the stripe and can be used to charge your account when you use it to buy goods or withdraw cash.

A development of these is the 'smart' card, which contains a computer processor capable of holding data about past transactions, and even updating the balance.

Voice recognition equipment

This is capable of recognizing the spoken word and transmitting it to the computer. It is becoming more widely available and its quality is greatly improved. An example of its use is in dictating a letter, which is then processed and printed out, or transmitted to another computer.

Computer-to-computer input

It is now possible to input data from one computer directly to another.

List as many examples of computer input devices as you can. Check your list with the above paragraphs - there are nine there altogether. What are the advantages and disadvantages of these three methods:

- keyboard
- bar code reader
- character reader.

In your workplace, what different methods do you use for inputting data to the computer? If your workplace does not use computers for data processing, what methods have you noticed other organizations using?

Hardware – output devices

These are used to transmit the resultant information to the user.

Printed output

Much output is printed or 'hard copy'. Some of this is necessary, for example when producing payslips for employees or when producing documents where the law insists on a hard copy. But it is true to say that a great deal of printed output is unnecessary, given the widespread use of computers. Much of the paper which is produced remains unused, especially where the information it provides is available in other ways, for example on screen.

Despite improvements in printer speeds, printing is slow and the cost of storage is high.

The screen (VDU)

The screen can be used in many situations as an alternative to printed output. Here are some examples:

- to check whether a product is in stock
- to check the current price or description of a product
- to check a customer's outstanding balance
- to view a list of invoices for the month.

Microfilm and microfiche

These are very similar in their use - one is in a roll and the other is a sheet. The computer can produce output directly in this form, which is then read by a special reader. The images on the form are minute and cannot be read by the naked eye. It is also a very useful method of storage for documents which are not required often.

Plotters

These are used to prints graphs and charts, and are particularly useful for technical and engineering drawings and plans.

Voice output

A common example of voice output is the British Telecom voice which tells you that 'the number you have dialled has not been recognized'.

Discs and tapes

Output onto these is really just a method of storage. The discs or tapes have to be read using the computer and screen.

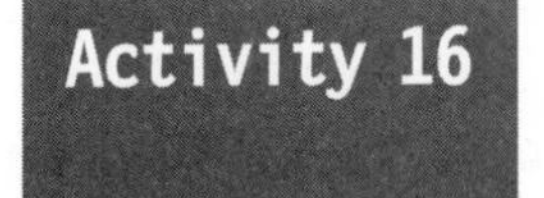

List as many output devices as you can. Check your list with the paragraphs above - there are six of them there. What are the advantages and disadvantages of using printers for computer output?

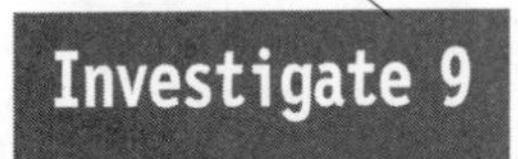

What devices does your organization use for computer output? If your organization does not use computers, what devices do other organizations use that you have experienced yourself?

Hardware – storage devices

Much of the data and information used by organizations needs to be stored for later retrieval, or to comply with the law. Storage is expensive, and hence computerized methods are extremely useful.

Some data and information is stored within the computer (on the 'hard' disc), but even so it is sensible to have additional storage as 'back-up' in case things go wrong. You will look at security aspects of computer systems in Chapter 5.

The most common storage device is magnetic disc. The small 'floppy discs' used by microcomputers are widely used. Magnetic tapes of various kinds and sizes are also used. Many microcomputer systems use 'tape streamers' to keep back-up copies of large quantities of data and information. The difficulty with a tape is that you cannot insert additional data without writing over the existing data and locating your data means running through the entire tape.

An increasingly used type of storage is the optical disc. Originally used only for reference material ('read only memory' or ROM), it can now be used to add data ('write' it) to the disc. A WORM (write once read many times) disc can be written to, and read many times, but the data cannot be removed. It is therefore a very secure method of storage and holds vast quantities.

CD-ROMs are also common, using audio technology to store data.

General purpose software

General purpose software is software that can be used in different ways by different users. Much of it nowadays comes as part of a 'package', including a word-processor, a spreadsheet, a database and perhaps graphics, with additional facilities as required.

The word-processor

This is a development of the typewriter, as a means of producing letters, reports, schedules, etc. The biggest difference, though, between the typewriter and the word-processor is the ability of the user to amend the document by inserting, deleting or changing items without the need to re-type the whole document.

Word-processors can be 'stand-alone' or 'dedicated' machines which only do word-processing. They have storage facilities, so that documents can be saved and retrieved later, and some have screens for viewing the document before printing.

Word-processors incorporated into computers can also interface with the other facilities of the computer to add diagrams, pictures, drawings, financial statements etc., or can pull out names and addresses and other data from a database.

The easiest way to understand how a word-processor works is to use one. Features include the following:

- adding, deleting or re-positioning individual or whole chunks of text
- choosing special styles of print
- incorporating charts, diagrams, pictures etc.
- automatic checks on spelling and grammar.

The spreadsheet

The spreadsheet is used primarily to perform calculations. It consists of rows and columns, like a table, and where a row and column meet is a 'cell'. The rows are numbered and the columns lettered, so where column M meets row 4 is cell M4. A cell can hold text or numeric data, including formulas. A typical spreadsheet has hundreds of columns and thousands of rows, making it possible to hold large amounts of data. Cells, whole rows or columns can be copied, moved, deleted, etc.

Several spreadsheets can be worked on at once with 'Windows' systems.

An example of the use of a spreadsheet is to project profit for the next 12 months, for an organization with three products. See Figure 4.2 for how the spreadsheet would look. Figure 4.3 shows the formulas that have been input to perform calculations automatically. In cell B7, the formula (B5 × B6) has been input to calculate the total revenue for product A.

	A	B	C	D	E
1	PROJECTED PROFIT FOR PRODUCTS A, B AND C				
2	FOR THE YEAR ENDED 30 SEPTEMBER 19X9				
3					
4		A	B	C	TOTAL
5	QUANTITY	1000	1500	1800	
6	UNIT SELLING PRICE	£10.00	£12.00	£8.00	
7	SALES VALUE	£10,000.00	£18,000.00	£14,400.00	£42,400.00
8	UNIT COST PRICE	£6.00	£8.00	£5.00	
9	TOTAL COST	£6,000.00	£12,000.00	£9,400.00	£27,000.00
10	PROFIT	£4,000.00	£6,000.00	£5,400.00	£15,400.00

Figure 4.2

	A	B	C	D	E
1	PROJECTED PROFIT FOR PRODUCTS A, B AND C				
2	FOR THE YEAR ENDED 30 SEPTEMBER 19X9				
3					
4		A	B	C	TOTAL
5	QUANTITY	1000	1500	1800	
6	UNIT SELLING PRICE	10	12	8	
7	SALES VALUE	+B5*B6	+C5*C6	+D5*D6	@SUM(B7 . . D7)
8	UNIT COST PRICE	6	8	5	
9	TOTAL COST	+B5*B8	+C5*C8	+D5*D8	@SUM(B9 . . D9)
10	PROFIT	+B7–B9	+C7–C9	+D7–D9	+E7–E9

Figure 4.3

A second spreadsheet can be set up to see what would happen if sales of product A increased by 10 per cent, and sales of product B decreased by 10 per cent. This is shown in Figure 4.4. No change has been made manually to any of the figures except the quantities of A and B - all the other calculations are performed automatically once the formulas are in place.

	A	B	C	D	E
1	PROJECTED PROFIT FOR PRODUCTS A, B AND C				
2	FOR THE YEAR ENDED 30 SEPTEMBER 19X9				
3					
4		A	B	C	TOTAL
5	QUANTITY	1100	1350	1800	
6	UNIT SELLING PRICE	£10.00	£12.00	£8.00	
7	SALES VALUE	£11,000.00	£16,200.00	£14,400.00	£41,600.00
8	UNIT COST PRICE	£6.00	£8.00	£5.00	
9	TOTAL COST	£6,600.00	£10,800.00	£9,000.00	£26,400.00
10	**PROFIT**	£4,400.00	£5,400.00	£5,400.00	£15,200.00

Figure 4.4

Spreadsheets can be used for many different purposes. It is the user who designs the spreadsheet for his or her own requirements.

Spreadsheets can be converted to diagrammatic form if required, using a simple charting option within the spreadsheet.

You can use a spreadsheet to help you answer many of the Activities and Case studies in *Resources Management*.

The database

A database is a file of data structured in such a way that a number of users can access it from a number of different points. The example quoted in many textbooks is of a telephone directory. The directory that most of you have in the office is a bound book. It is definitely NOT a database. If you want to know a number, you need the person's name and possibly their address too. If you do not have the name, but only have an address, it is virtually impossible to locate the number. And if you only have the number - it would be a very long search indeed to find the name of the person.

A database would allow you to access the records from any of these starting points, and even perhaps with only part of a number. A telephone number might be 012928 515063. If you knew the surname was 'Ogden' and the last part of the number was 515063, the database would list all the Ogdens with that number, and give you the code and the address for you to confirm.

Databases can hold vast amounts of data, and if they are properly constructed, can be extremely flexible. Because of the ability to access the data from any point, they get over the problem we looked at in 'other data' on page 37 of several users keeping individual copies of the same data but in different orders.

Graphics

A graphics package can be used to greatly enhance the presentation of reports etc. It incorporates pictures, backgrounds, titles and headings, bullet points, charts, diagrams and so on. They are generally very easy to use, with instructions clearly given in the package.

They can be used to produce reports, slides and to give moving presentations. They can even incorporate sound, photographs and films, with additional equipment, and add a touch of professionalism to the provision of information.

Desk-top publishing

A combination of all the above components of a general-purpose package produces a system which is capable of providing very high-quality leaflets, publicity materials, brochures, etc., which previously could only be produced by specialist printing firms.

What are the five different types of general-purpose software? Why is it called 'general purpose' software? Which piece of software would you use in each of the following situations:

- producing a cash budget for the next 12 months
- producing a report on the problems of maintaining production machinery
- maintaining details of customers
- producing an advertising leaflet
- making a presentation about the company to a group of customers.

See Feedback section for answer to this activity.

Application software

This is software designed for a specific use. It can be written for you by your own computing staff, outside experts, or bought 'off the shelf' as a ready-made program.

As with general-purpose software, off-the-shelf applications often come as a package of several elements together. The advantage of this is that they will interlink with one another.

Examples of application software are systems for:

- financial accounting
- payroll
- stock control
- purchasing
- invoicing
- costing
- personnel records.

Systems such as these are often *menu-driven*. This means that to use them, you choose from different options (or *menus*) displayed on the screen. Having made your choice, another menu appears with further choices. The list above could be the first menu in a system. Let us say you choose 'financial accounting'. The second screen might show:

- sales ledger
- purchase ledger
- nominal ledger
- exit to previous menu.

Choose sales ledger and the third screen might show

- input transactions
- update customer file
- produce reports
- exit to previous menu.

Choose update customer file and the fourth screen might show

- add new customer
- delete customer
- amend customer details
- exit to previous menu.

If purchased as a package, then the different systems could interlink. If goods are out of stock, the stock control system could cause the purchasing system to raise a purchase order. When the invoice arrives, it could be used to update the purchasing records, the stock control records and the accounting records with just one input.

These systems often provide comprehensive reporting facilities too, so the purchasing system could produce a list of outstanding orders. Some of these reports can be tailor-made by the user to suit their own purposes.

Activity 18

Identify four types of application software.

See Feedback section for answer to this activity.

Methods of input, processing and output

Methods of input

There are two methods of inputting data to the computer. Items can be input singly, i.e. one at a time, as they arise. Or they can be input in groups (called *batches)* of the same item together.

The advantages of inputting in a batch are:

- the operator can build up speed as he or she becomes familiar with the data to be input
- the batch can be totalled in advance, to ensure that all items are input. The totals can include the total number of items, the total value, or even the total of the document numbers (e.g. invoice numbers) to act as an additional check. This latter total is called a *hash total*.

The disadvantage of batch input is that you have to wait until you have a number of items to input together, which might mean that the files are not always up-to-date.

Methods of processing

There are also two main methods of processing data. These are *real-time processing* and *batch processing*.

Real time processing is where items are input and the files are updated immediately. This is invaluable where files need to be as up-to-date as possible. Out-of-date files can result in wrong decisions being made.

Batch processing is where a number of documents or items of data are processed together, in a batch. The items need not have been input as a batch, they might have been input singly and stored in a transaction file. As with batch input, the files might be out-of-date for most of the time. This is not always a problem. Batch processing is ideal for updating the payroll, as this only needs to be done once a week or once a month. But real-time processing is essential for updating stock records as incorrect records could lead to customers being given wrong information, or to a shortage of stock.

Methods of output

The method of output depends primarily on how quickly the information needs to be accessed. Batch output is usually reserved for periodic reports, such as a list of invoices issued during the year, perhaps output on microfilm, or a printout of unfufilled orders during the month. For quick decision-making, however, output needs to be immediate, and therefore instant access to the master files is needed.

In such cases, access to the files needs to be *on-line*, which means the output device needs to be connected directly to the files. A good example of this is when a customer needs to know if a particular item is in stock. The operator can access the files directly and is able to give the information at once. Another example is a customer requesting goods on credit – the order department needs to know whether or not the customer has exceeded their credit limit.

For on-line access to be useful, files must be updated in *real-time*, rather than by batch, otherwise they will be out-of-date.

Briefly explain the following terms:

- batch input
- batch processing
- real-time processing
- on-line output.

Refer to page 51–2 above to check your answers.

Summary

In this chapter you have looked at the methods of recording and storing data in both manual and computerized systems. You have learnt the difference between 'hardware' and 'software', and about the various files used in processing data. You have looked at a range of items of hardware for different purposes, and their advantages and disadvantages. Software for both general purposes and specific applications has also been described.

If your own organization is not yet computerized, you should now be better placed to understand how computers could be used to perform the various tasks which are currently undertaken manually. You will have learnt some useful terms to enable you to communicate with computer-based staff and suppliers of computer systems, more easily.

Review and discussion questions

1. What are the component parts of a computer system?
2. What is the difference between 'hardware' and 'software'?
3. What are the four files used by computer systems?
4. What is a program?
5. Which input devices are suitable for the fast input of data which needs to be highly accurate?
6. What types of output device are there?
7. Name three storage devices in computer systems?
8. List some of the features of a word-processor.
9. What is a cell in a spreadsheet?
10. What is the main difference between a database and a file of data?
11. What is 'application software'?
12. What methods of input, processing and output are there?

Case study

Georgina and Robert manage a small nursing home for the elderly. There are 20 residents at any one time, staying for varying periods from one week to permanent residence. They are responsible for the following functions:

- maintaining records of residents, their medical details, their relatives
- billing residents and the local authority monthly and recording payments received
- ordering food supplies
- calculating wages for six full-time and eight part-time staff
- ordering and maintaining records of drugs in stock and used
- producing quarterly reports on income and expenditure
- producing brochures on the facilities of the home for advertising purposes and for new residents and their families.

Produce a short report for Georgina and Robert, suggesting ways in which a computer could be used to assist in the maintenance of the above systems. Consider the hardware needed for input, output and storage. Which systems would need on-line access? What types of software would you recommend?

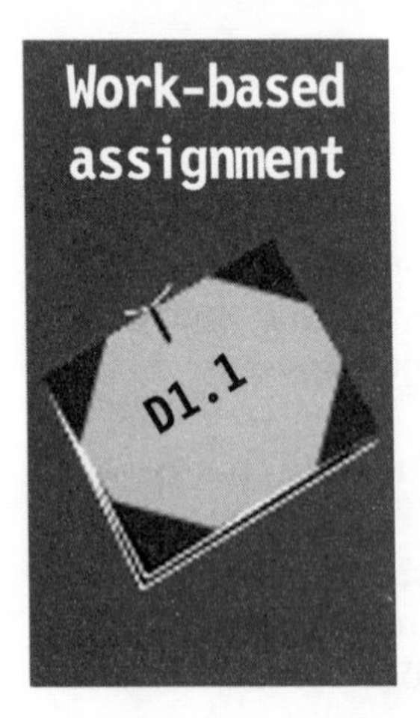

Select a computer system at your workplace and identify the following details about it:

- the hardware used, for input, processing and output
- the software used
- the methods of input, processing and output
- any areas of the above which could be improved.

If your organization does not have computer systems, select an area which you consider might benefit from computerization and make suggestions for hardware, software and methods of input, processing and output which might be appropriate.

5 The use of information systems

Learning objectives

On completion of this chapter, you will be able to:

- explain what is meant by an *information system*
- describe the different types of information system
- understand the importance of technology in communicating information
- explain the advantages of technology in providing information
- understand the effects of technology on users
- identify areas where users can be involved in the design of information systems
- appreciate the need for, and methods of security in information systems.

Introduction

Computers are now widely used in the provision of information. The term *information system* refers to any system which is used to provide information. A *management information system* provides information to management at all levels.

Because managers need different types of information, there are many different systems available to meet their needs. Technology contributes towards those needs. This chapter looks at some of these systems.

Types of information system

Transaction processing systems

These include the systems used in the processing of financial transactions, such as:

- accounting systems
- payroll systems
- stock control systems
- costing systems

which are used to produce the organization's accounts, budgets and to perform essential day-to-day tasks.

Reporting systems

These are systems which use transaction processing systems to produce reports, either on a regular or an ad hoc (i.e. when required) basis. A regular report might be a monthly listing of all debtors outstanding for more than 1 month. An ad hoc report might be of customer orders for a particular product to determine the pattern of customer demand over a period.

These systems are sometimes referred to simply as *management information systems,* because they provide basic information to management from the normal transaction processing systems. They are usually required by operational level management, as their results are based on past data.

Decision support systems (DSSs)

These are an extension of spreadsheets, but use complicated mathematical formulae to enable less structured decisions to be made. Some can be purchased with the formulae already included. The user needs to be reasonably computer-literate to be able to use them fully.

The idea of DSSs is that they can be used to produce a variety of alternative solutions to problems. They do not make the final decision however - that still rests with the management.

Executive information systems

As the name suggests, these are used by top-level management. They are very powerful, often linking in to all of the organization's internal systems, and even to external systems such as commercial databases. They enable the executive to make decisions involving complex, unstructured problems.

However, because many executives are not computer-literate, these systems must be very easy to use. They must also be capable of producing output in many different formats, for presentation to members of the board, major customers, etc.

Expert systems

These are designed to be used instead of human experts in particular fields. They contain a vast amount of rules and assumptions, such as those you might expect a human expert to have. For example, a medical expert system would include factual medical data, details of drugs and their side effects, and could be used to suggest treatments for patients. As each patient receives the treatment, the effects are fed back into the system which is said to *learn* from the results by updating its bank of knowledge. When it is next used for the same problem, it might produce a different treatment as a result of what it has learnt from its previous attempts.

An expert system needs to be operated by someone with some knowledge of the area it covers.

Design systems

There are also systems used by specialist designers of machinery, office layouts, computer programs, buildings, etc. They can be used to provide diagrams, often three-dimensional, which the designer can use to build models. They can also perform mathematical calculations to calculate quantities of materials and so on.

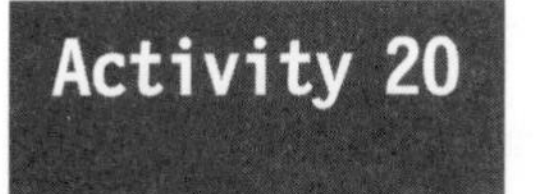

Name six types of information system and give a brief outline of what each one does.

See Feedback section for answer to this activity.

What information systems do you have in your own workplace? If your organization is not heavily computerized you might have only transaction processing systems - and they may be manual. How could you envisage computerized systems being of use to your organization?

Communication systems

The use of electronic communication systems is growing very rapidly indeed. The need for large numbers of people to have access to information very swiftly, and to be able to communicate it to others, is very important. New systems are being developed weekly which provide new methods of doing this.

Terminals

These are pieces of hardware which are situated away from the main computer processor, but allow the operator to enter data into, and obtain information from, the computer. Strictly speaking, a terminal does not do any processing itself, this is done by the central computer. The terminal and computer merely communicate with each other.

Electronic point of sale (EPOS) terminals

These have become commonplace in retail outlets in recent years and in other areas too. Where there used to be a cash-till, this has been replaced by the EPOS terminal. The terminal is connected to the computer and acts both as an input and an output device. The bar-code scanner inputs data to the computer regarding items sold, and the computer relays back the price and description. Stock files and other files are immediately updated. It follows that EPOS terminals are on-line, and processing is real-time.

Electronic funds transfer (EFT)

This is a system which allows a bank to transfer amounts from one person's account to another person's account at another bank. There are various different types of EFT.

Electronic funds transfer at point of sale

This is a development of EFT and EPOS where customers can pay at the point of sale, and the money is transferred automatically from the buyer's to the seller's bank account.

Electronic data interchange (EDI)

This enables the computers of two different organizations to interact with each other. A common use of EDI is where an organization's stock falls to the level at which they need to re-order. Rather than raising a manual order form, or making a telephone call, the supplier's computer system is automatically notified of the requirements, and can respond with a delivery date and other details back to the buyer's computer system. All communication, including the delivery documentation, the invoice and payment, can be carried out electronically.

The two computer systems must be compatible and many large organizations have forced out smaller suppliers who could not afford to purchase the necessary equipment. The system does mean that paperwork is almost eliminated and that goods can be ordered and delivered much more quickly. As a result, a system called Just In Time (JIT) has evolved, whereby organizations can keep their stocks down to a very low level, knowing that new supplies can be obtained very rapidly.

Networks

A network exists where one or more computers are connected to each other, or to a central computer. This means that data and information can be passed around. Processing can take place either on the central computer or on the individual computers.

Local area networks (LANs)

A LAN exists where computers are connected via cables, usually within a building or a relatively small area.

Wide area networks (WANs)

These use the telephone system to connect users over a wider area, even across the world.

The Internet

This is an international communication system which enables users to access information from around the world. The information held on the Internet is dependent on what organizations choose to display, but there is a growing wealth of marketing information, financial data and topics of general interest. People can buy and sell goods and services through the Internet. Users are able to link up with other users via a variety of 'clubs' and 'societies'. It is regarded as a possible medium for business users to communicate with each other on all manner of subjects. The 'net' uses text, pictures, animation, film, sound, etc., to display its information.

The Internet is also known as the World Wide Web, as its design is rather like a spider's web with every point on it being connected to all the others.

Mailing systems

Sending messages via computer can be very time-effective. Imagine the following scenario. You need to contact your boss who is in Rome. You phone him during the morning, but he is out at a meeting. He returns your call in the afternoon, but you are out of the office. Imagine if he were in America where the office hours do not coincide with those in Britain.

An electronic-mail (E-mail) system allows you to transmit a message to his computer, where he can pick it up at his convenience, and transmit a reply to you.

A more recent development is *voice-mail*, whereby a verbal message can be recorded and picked up at the recipient's convenience.

Teleconferencing systems

These enable users at different locations to 'converse' with each other - and even several people at a time - using their computers. With the addition of video and voice transmission, these conversations become close to 'real-life' conferences.

The great advantage of teleconferencing is that employees do not have to travel far in order to conduct such meetings, with an obvious reduction in time and cost to the organizations involved.

Telecommuting

The use of networked computers means that employees can work from home rather than travelling into the office. By having access to central files and electronic transmission of data, all that they need to perform their work is accessible. It can be used by organizations employing home-workers on a permanent basis and means that they do not have to provide office facilities, but it can also mean a lack of control over the employees' work. Employees, too, may miss out on the interaction with colleagues. It can also be used by senior executives who often work partly at home and partly in the office, saving travelling time and cost.

Name six types of electronic communication system. Give a brief description of each one, and how it might be used in an organization.

Consult the paragraphs above to compare your answer.

Investigate 11

What types of electronic communication system are in use in your own workplace? How easy are they to use? Which ones do you use yourself? Which are the most useful and why? Are any of the systems under-used, and what are the reasons for their under-use?

The effects of technology on users

Introducing change

Human beings are, by nature, afraid of change and the introduction of technology is no exception to that. Employees may be afraid of losing their jobs, of being unable to cope with the changes, or of losing their personal identities in the workplace. It is important that any change is managed sympathetically, with users in mind. Some tactics for a smooth introduction are:

- involve users at all stages, not just at the end when the decisions have all been made
- ask users' advice on changes to be made
- give users 'ownership' of particular areas to be investigated and developed
- emphasize the benefits of changes, such as improved job satisfaction, the acquisition of new skills
- introduce changes gradually
- support users throughout, by encouraging them to voice concerns and by responding sympathetically
- ensure that proper is training is given to users
- avoid unnecessary break-up of existing successful working relationships.

Human–computer interfaces (HCIs)

It is important that users are comfortable with using computers. The computer is said to *interface* between the user and the system. That interface should be as 'user-friendly' as possible. This means that it should be easy to use, difficult to cause damage, and reliable.

Windows environment

Many modern systems operate on a system called *Windows.* This presents the user with an easy-to-understand screen of options and choices, using abbreviations, pictures and instructions where necessary. It is called *Windows* because it is possible to switch from one screen to another to access other information or instructions. For example, in order to print a document, the user can locate a picture of a printer on the screen; to check spelling, the picture shows the letters *ABC.* These pictures are called *icons.*

Within the various choices shown on the screen, there are additional choices which can be accessed by the click of the mouse, or by the keyboard. For example, if the user chooses `Text`, a further list of choices showing typefaces, font size, character styles, etc., is shown.

User intructions

Many packages also include user instructions and 'help' screens to assist in performing unfamiliar tasks. There are even inbuilt tutorials in some packages, for users to teach themselves.

End-user computing

With the availability of user-friendly systems, many users are learning to develop their own systems for particular purposes. These range from designing documents and building spreadsheets, to producing their own reports from existing systems and even designing new systems.

It is important that users are properly trained and so some organizations have set up *Information Centres* to control the work of end-users. Users are required to register any reports, systems, etc., that they have developed, to ensure that they conform to company guidelines (especially where they are being sent to customers outside the organization) and to ensure that work is not being duplicated elsewhere. The information centre also offers support and training where required.

The encouragement of end-user computing relieves the pressure on specialist computer staff, who are then available to develop more complex systems. It also means that users are in charge of their own work, which can improve motivation.

If you were changing your existing manual system for a computerized system, what factors would you take into account to ensure a smooth changeover?

See Feedback section for answer to this activity.

The advantages of computer technology in organizations

There are many advantages of using computer technology. These include:

- speed – of input, processing and output
- accuracy – of input, processing and output
- the ability to process large volumes of data at once
- flexibility – in communication and presentation of information
- accessibility of data and information to a variety of users
- the ability to store large volumes of data and information in a small space.

Security in computer systems

Security is vital in all systems but especially in computer systems when there is so much information held in them and they are so easily accessible by a number of people.

Breaches in security can result in several problems:

- loss of data or files
- inaccurate data or files
- access to confidential data.

Breaches in security can be either *accidental* or *deliberate*.

Accidental breaches in security

Inaccurate data input

Some methods of data input are more accurate than others. Those which involve the input of data via a keyboard are very prone to errors. To overcome this problem, data can be input twice and the two files checked against each other. This is time-consuming and costly. Such a check is called a *verification check*.

An alternative is to check the screen as data is input and a number of errors can be spotted visually. Some, however, will slip through unnoticed.

The computer itself can be programmed to check for some errors. For example, the date on an invoice of 31/09/98 can be identified as incorrect as September has only 30 days. The computer screen can highlight this, allowing the operator to check again and correct the input. Checks like this are called *validation checks*.

The computer can also check that the customer number on an invoice is correct by comparing it with the customer master file. If that customer does not exist on the master file, the number might be incorrect. It could, of course, be for a new customer who has not yet been recorded on the computer. This is another type of validation check.

Inaccurate processing

The computer should be programmed to perform its tasks accurately. If it does not, then the output will be incorrect. This can only be avoided by ensuring that the program is properly written and tested when the system is installed. Testing involves feeding in a number of test data, of all kinds, both correct and incorrect, and checking what the computer does with the data. Testing is a vital part of system design and implementation.

Deliberate breaches of security

These fall into three categories.

Fraud

This is where someone tampers with the system in order to steal something from the organization. Examples include computer programmers writing a program which miscalculates wages by a penny per employee, putting the pennies into an account in their own name and transferring the funds to their own bank account. Another example is an employee setting up a bogus supplier on the purchase ledger, submitting a false invoice and obtaining payment.

Another kind of fraud is where a person takes a copy of an organization's files for their own use. A recent case occurred where an employee printed out the details of the company's new product designs and sold them to a rival company.

Hacking

Individuals who enter a computer system without authority, usually from another computer, are called *hackers*. Often they do not intend to cause harm, but merely to prove that they can do it! The main problem is that this invades the privacy of the company and the individuals whose data are stored on their files.

Viruses

These are instructions fed into computer files or programs which corrupt the data stored on them. They are called viruses because they enter the system from other computers which already have the problem and they spread like a common cold. There are thousands of different types of virus and they enter the system by using discs which are already infected, or by the user copying software that is illegal. Most software is protected by copyright and some writers include viruses which can detect illegal copying.

Dealing with breaches of security

The five aims of security are to:

- prevent
- detect
- recover
- correct
- deter.

Prevention

Prevention is the first aim. It can be achieved in a variety of ways, such as

- having adequate locks on rooms
- restricting access to those with keys or entry-passes
- fastening equipment firmly to walls and structures
- using buzzers and alarms
- introducing passwords and other access controls so that individuals can only access data and files which they are authorized to access
- having separate passwords for those who are allowed to alter data
- dividing duties so that no one individual is able to control a whole area of activity, e.g. the programmer should not be allowed to set up new suppliers; the cashier should not be allowed to input invoices, etc.
- proper training of staff
- careful recruitment and selection of staff
- adequate pay scales for employees, to avoid staff resenting their employers.

A large number of breaches of security are carried out by disgruntled staff.

Detection

Discovering a breach can be difficult, and may not occur until some time after the actual breach has been committed. Often it is discovered only when something noticeable has occurred. The fraud of transmitting odd pence into a separate account would not be noticed on a weekly basis. Only when a large cheque is made out for the total would it be realized.

Possible methods of detection might be:

- regular checks on input and output
- keeping a log of users of the system (perhaps by computer)
- having regular audits and spot checks.

Recovery

Having discovered the breach, it is necessary to recover the lost data or correct the errors. It is vital that copies are kept of all data, files and programs, away from the normal storage area. It is also a good idea to keep more than 1 month's files, so that, if necessary, you can reconstruct files from some time ago and bring them up to date.

Correction

The errors must be corrected at once. This might mean installing additional security measures as well as correcting files and programs.

Deterrence

It might be necessary to implement penalties for breaches of security, from loss of wages to loss of job for certain serious offences.

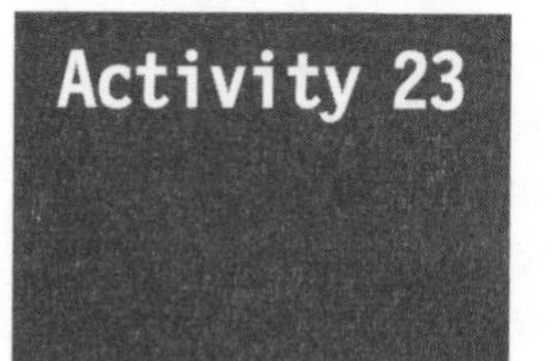

You are installing a new computer system to process the payroll. Identify as many possible sources of danger to the security of the new system and the ways in which you could avoid breaches of security.

See Feedback section for answer to this activity.

Summary

In this chapter you have looked at the different types of information system for different purposes. You have looked at the developments in technology in providing information to different groups of people and by different methods. You have also learnt how changes in the working environment, including the introduction of new technology, can have an effect on the workforce and you have seen how important it is to introduce changes properly.

Finally, you have seen how important security is in information systems, and have considered ways in which security can be maintained and improved.

Review and discussion questions

1 Explain what is meant by an *information system*.
2 Describe the different types of information system.
3 What are the advantages of computer technology in information systems?
4 What is meant by a 'network'? What are the different types of network?

5 How is it best to deal with the human aspects of changes in technology?
6 What are the five aims of security?

Case study

Wing is a manager in a Citizens Advice Bureau (CAB). The Bureau gives advice to the public on a variety of legal matters, including entitlement to state benefits, how to complain about faulty goods and services, how to take out legal action against someone, what to do if a neighbour is causing a nuisance and so on. The list is almost endless. The Bureau is staffed by three paid employees who have experience in the various areas mentioned and there is access to a number of unpaid volunteers who assist in form-filling and contacting outside organizations.

The staff frequently telephone other CAB offices to ask for advice and information, but so often the lines are engaged. Sometimes there is a need for a client to discuss his or her problem with a number of experts at the same time.

A client comes in one day to ask for advice regarding a settee he bought 4 months ago. Within days of receiving the settee, he noticed faults in it which he notified to the suppliers at once. The company involved have still done nothing to rectify the faults, with a series of excuses. The client now wants to know what legal action he can take.

Wing's staff are unsure about the advice they should give. The bureau has dealt with many similar cases before, but the staff who dealt with them have since left. The solicitor who usually offers help is on holiday.

Wing is keen to help this client. The computer system, which contains a database of previous similar cases, is unreliable because someone has been deleting records of past cases, thinking that they would no longer be needed. As Wing is searching through the computerized list of solicitors, the system fails and produces a message that 'the system has been corrupted. A virus has been detected'.

Advise Wing on the systems that he could use to help this client. What could have been done to ensure that the database was accurate? How can Wing get round the problem of the solicitor being on holiday? If he cannot get through to other CABs on the phone, how else could he contact them? What could he have done to prevent the virus from attacking the system?

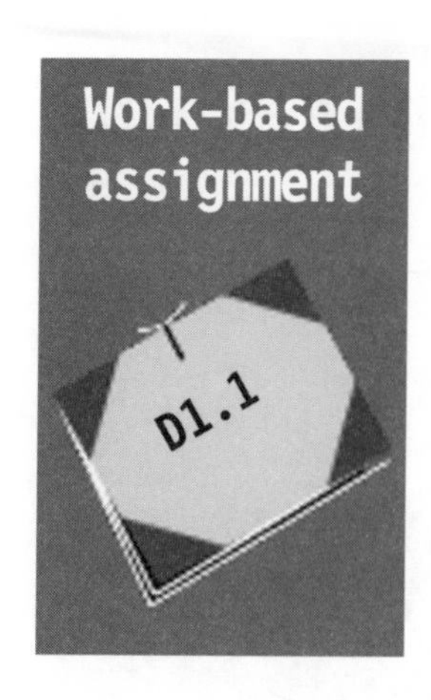

Investigate the types of information system in use at your workplace. Produce an informal report covering the following areas:

- the types of information system used
- the methods of communication used
- the security procedures in place.

In particular, identify any problems that you are aware of, which have occurred in the past year, and give details of how they were overcome.

6 An introduction to communication

Learning objectives

On completion of this chapter, you will be able to:

- explain why communication is important
- differentiate between and describe, oral, written and visual communication
- describe the advantages and disadvantages of the different types of communication
- explain the model of the communication process
- describe the barriers to communication
- suggest methods of overcoming the barriers to communication.

Introduction

It is essential that both you and your organization can communicate effectively. It does not matter how good you are at the technical aspects of your job, if you are not able to communicate effectively both by the spoken word and the written word you will be unable to communicate your ideas and instructions to others. The ability to communicate effectively is an essential skill for a successful manager. Communication is the process of creating a common understanding, interpreting ideas, opinions and feelings, between two or more people. In this chapter we look at the different types of communication and the communication model.

The importance of communication

People spend nearly 70 per cent of their waking hours communicating - writing, reading, speaking and listening. An idea is useless until it is transmitted and understood by others; good communication within your workteam is essential for the team to function effectively. We look at communication and your workteam in Chapter 9 of this book. It is also vital that there is good communication throughout the organization: communication in organizations is also examined in Chapter 9.

Communication is used at work, to provide people with the information they need to make decisions, to motivate people by explaining what needs to be done, setting goals and providing feedback. Communication is also important to people socially; we all have social needs that need to be met at work.

Types of communication

There are various types of communication, all of which are used in organizations and all of which have their advantages and disadvantages. Whatever type of communication you use remember the five C's of communication: ensure that all your communications are:

- **C**lear
- **C**omplete
- **C**oncise
- **C**orrect
- **C**ourteous.

Oral communication

Examples of oral communication are face-to-face conversations, meetings, presentations, interviews, training and talking on the telephone. The advantages of oral communication are that the sender gets immediate feedback and it is much easier to communicate attitudes and feelings. However, oral communication makes it more difficult to think things out because you have to respond very quickly. There is no written record of what has been said, which can lead to disputes later, it can be difficult to hold your ground in the face of opposition and it is more difficult to control when a number of people are involved.

Oral communication is reinforced by *non-verbal behaviour*. It is not only what you say that is important, but the non-verbal messages that accompany the words. Only a small percentage of the impression you make on other people stems from verbal communication. What makes a much greater impact are the non-verbal messages that accompany the words. Such as:

- vocal pitch and emphasis
- speed of speech
- clothing/dress
- body language, which is about your breathing, posture, facial expressions, eye contact, gestures and movements.

We will look at oral communication, listening and body language in more detail in Chapter 7.

Written communication

Examples of written communication are letters, memos, notices, in-house publications, manuals, handbooks, reports, minutes of meetings and computer output. The advantages of using written communication are that the recipient can read the message when they choose to, written communications can carry complex information, they can be widely circulated, they provide a permanent record and, if delivering a 'difficult' message, will evoke fewer emotional responses as the recipient has time to absorb and consider the message. However, written communications can take time to produce, they tend to be formal and distant and can cause problems with interpretation. Instant feedback is not possible and once you have committed yourself to the written word it can be difficult to change the message. Written communication does not allow for a rapid exchange of opinions, views or attitudes. We will examine written communication in more detail in Chapter 8.

Investigate 12

What methods of written communication are used by the senior managers in your organization to communicate with the staff? What is your opinion of the quality of communication within your organization?

Visual communication

Examples of visual communication are diagrams, charts, tables, slides, films, videos, models. Visual communication methods are often used to support oral communication and the written word. They provide a visual record, visual stimulus and can help to simplify difficult concepts. They can simplify messages with numbers in them and be used to illustrate techniques and procedures. The disadvantages of visual methods are that they may be difficult to interpret without the reinforcement of the written and spoken word, they may be costly and time-consuming to produce and distribute and more difficult to store.

Draw up a simple table like the one below, and make a note of the methods of communication used by you during a single day at work, to both give and receive information. Put a tick against each one every time it occurs. Which ones are most frequently used?

Method of communication	To give information	To receive information
Oral		
Written		
Visual		

A model of communication

A model of communication is shown in Figure 6.1. It shows that before communication can take place an idea expressed as a message is needed. It passes between the sender and a receiver. The message is encoded (devised) by the sender and passed by way of some medium to the receiver, who

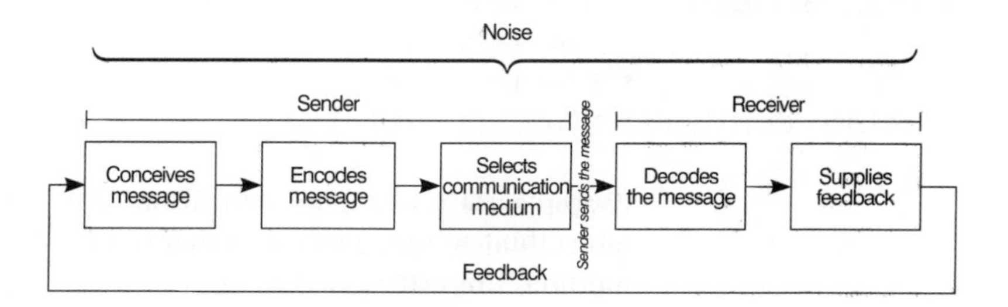

Figure 6.1
A communication model

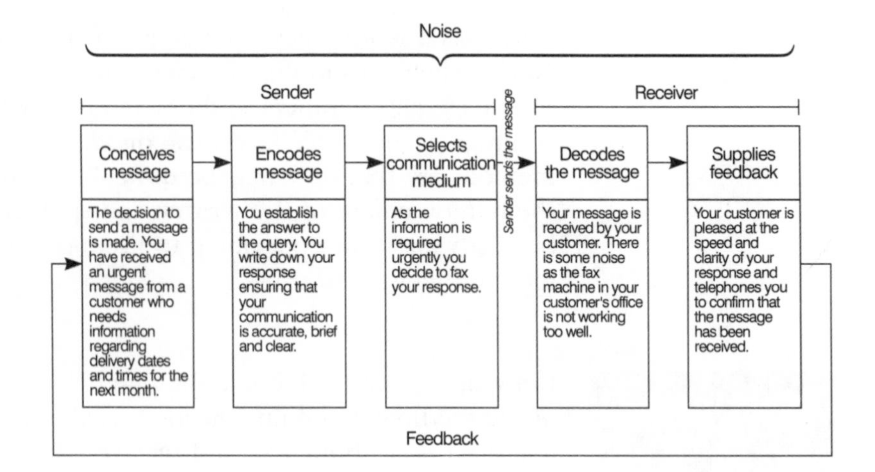

Figure 6.2
An example of the communication process using the communication model

retranslates (decodes) the message. The noise that is shown on the communication model is anything that gets in the way of the message which might distort its meaning. The result of the entire communication process is the transfer of meaning from one person to another.

Figure 6.2 shows an example of the communication process, using the model.

We will now examine the communication model in more detail.

The sender conceives the message

The sender makes a decision that a message needs to be sent. This might be prompted by a variety of things, for example, an idea of their own, an incoming telephone call or letter, a conversation, or issuing a communication might be a stage in a procedure that has to be carried out.

As a team leader you will often have to make decisions about exactly what information to pass on to another person or people. You have to decide:

- what people *must* know
- what people *should* know
- what people *could* know.

It is important that your workteam have sufficient information to do their job properly and that they are not burdened with too much information.

The sender encodes the message

You want to send a message. You have to select the appropriate language to communicate the message clearly. You may choose to communicate a message visually (such as a No Smoking sign), orally or in writing. Whatever the method you choose, it is useful to remember the ABC of communication:

- **A**ccuracy
- **B**revity
- **C**larity.

It is important when you are encoding a message to remember who the receiver is. For example, it would be wrong to use technical jargon, which is only understood in your own organization, in a letter to a customer. Just as it would be wrong to explain in detail a technical term that everyone within the organization understood, if you were writing an internal memo.

The sender selects the medium

Now you have to select the medium in which to send the message.

Activity 24

You may decide to send a message via a written medium. List five examples of a written medium you could use.

See Feedback section for answer to this activity.

Activity 25

List three advantages of written communication.

See Feedback section for answer to this activity.

What method would you use to send each of the following messages:

1 Send a copy of a two-page document to Head Office, that the marketing director requires urgently.
2 Ensure that all your workteam know how to use the new photocopier.
3 Inform your workteam about a new product that your organization is introducing.
4 Apologise to a customer about a mistake that has been made that has adversely affected the customer.

See Feedback section for answer to this activity.

Various factors have to be taken into account when you are selecting an appropriate medium for a message, for example:

- *cost* - a courier is expensive
- *speed* - E-mail and fax are fast
- *accuracy* - written communication is more accurate than oral communication
- *the nature of the message* - a fax might not be confidential
- *the nature of the receiver/s* - a telephone call might be too casual a way to respond to a customer's letter
- *scale of the task* - it is usually easier to write or E-mail a large number of recipients
- *importance of feedback* - oral communication is better for immediate feedback.

The receiver

The receiver's role in communication is absolutely crucial. Good communicators are not just good at sending messages, they are also good at receiving messages. This means carefully listening to oral communication and carefully reading written communication, before responding in an appropriate way.

The receiver decodes the message

When the receiver gets the message, they decode it. The process of decoding can be difficult, especially if there is a lot of *noise* surrounding the message. Noise, in relation to the communication model, is anything that gets in the way of the message. It might actually be talking in a noisy environment or the fact that the telephone or computer is not functioning properly. It could be that the receiver is hungry and is thinking about how near lunch time it is. It could be that jargon is being used that the receiver does not understand or perhaps the receiver is thinking what dreadful socks the sender is wearing and that they do not match the rest of their clothes.

The receiver provides feedback

Following receipt of the message the receiver provides feedback; the type of feedback will depend on the message. It might be confirming that the message has been understood, confirming that appropriate action will be taken, seeking further clarification or whatever feedback is appropriate in the circumstances.

Activity 27

Use a message that you have communicated today and complete the boxes in the model of the communication process below in relation to your message.

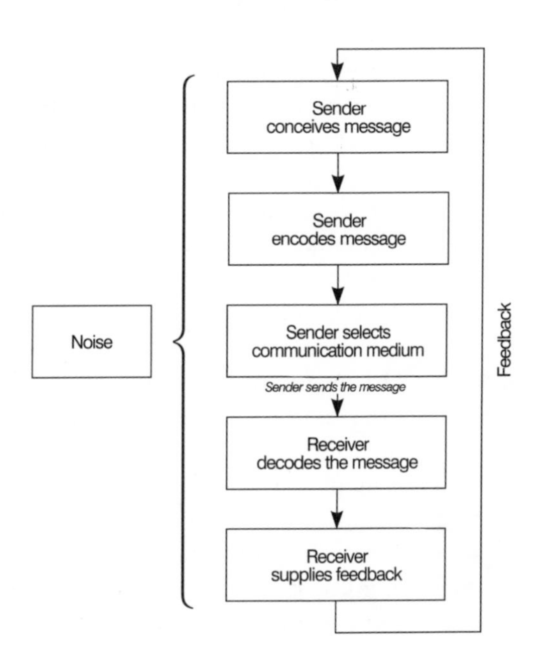

Barriers to communication

There are many barriers to good communication which you need to be aware of and consider in all your communications. If you do this it will ensure that all your communications are effective.

- *communicating too much* – information overload
- *encoding or decoding the message incorrectly* – so that a misunderstanding occurs
- *choosing an unsuitable medium*
- *failure to feedback* – if the message is not confirmed it may have been unknowingly misunderstood
- *incorrect vocabulary* – for example, jargon to customers
- *incorrect or incomplete information*
- *differences in people* may mean that they interpret messages in different ways – for example, differences in age, sex, culture, background, education and language
- *status differences* – people may not listen as carefully to a message from a subordinate as to one from their manager
- *conflict between individuals* – if you do not like the person who is delivering the message, it may affect the way in which you interpret the message
- *individual bias and selectivity* – that is, we hear what we want to hear, see what we want to see
- *verbal difficulties* – an inability to articulate the message clearly
- *lack of practice in written communication skills* – resulting in a poorly constructed written message
- *fear and emotional overtones* – can cloud the message; if a person has bad news to pass on, the sender may tend to avoid the truth and not pass on the whole message.

List as many occasions as you can where an item of information has been communicated to you and you have not fully understood it. Do the same for information which you have communicated to someone else and which was not fully understood. For each occasion, was there the opportunity for feedback? And could you suggest how the communication could have been improved?

Breaking the barriers

Now you are aware of the communication barriers, it is important that you make an effort to break them. Following the guidelines below will help you:

- make sure that your communication has a clear aim
- take time to prepare the communication
- choose the right setting for oral communication
- ask for help and advice, if you need it
- anticipate any queries the recipient might have
- always select the correct medium
- construct your message bearing in mind the recipient/s
- check understanding by ensuring you get feedback
- take the opportunity to undertake training in communication skills
- remember the ABC and the five C's of communication.

Activity 28

For the message that you used in Activity 27, what were the potential barriers to communication and how did you overcome them?

Summary

This chapter has introduced the various types of communication and the model of communication. You will now be able to consider more carefully the way in which you construct your communications and be able to select the appropriate medium for your message. We have touched on many aspects of communication such as oral and written communication, communication within your organization and workteam, which we will explore in more detail throughout this book.

Review and discussion questions

1. Define communication.
2. What percentage of their waking hours do individuals spend communicating?
3. Why is communication important in the workplace?
4. List the five C's of communication.
5. Give three examples of oral communication.
6. List the disadvantages of written communication.
7. What are the disadvantages of visual communication?

8 Explain what is meant by the term 'non-verbal behaviour'.
9 List five of the factors that you should take into account when selecting a medium for a message.
10 Recall an occasion when a communication went wrong, for example, you or a member of your team wrongly interpreted a message. What were the consequences of the poor communication. Why did it happen? What have you learned from the experience? How will you make sure that the same mistake does not occur again?

Case study

Lydia is 21 years old, she is white and was educated at a private school. Lydia is single and loves pop music, going to see bands and dancing the night away at night clubs. She likes to keep fit and particularly enjoys swimming.

Rashid is Susan's line manager. He is 39 years old and was born in Pakistan, where he lived for the first 12 years of his life before moving to the UK and attending an inner-city comprehensive school. Rashid speaks English well, although he has quite a pronounced foreign accent. Rashid is married with four young children. He enjoys tennis and is passionate about cricket.

What possible barriers to communication might exist between Susan and Rashid?

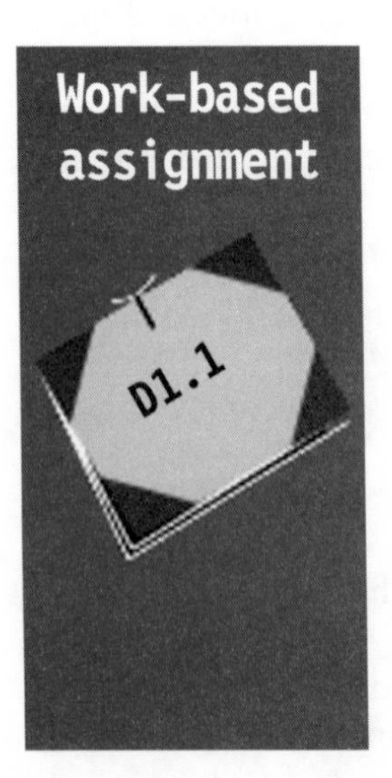

Analyse your communication experiences for one day at work. Record every communication contact that you have during the day. Calculate what percentage of your time in total you spend communicating. How much of that time are you communicating orally?

Select ten communications that you have received during the day. Compile a chart similar to the one shown overleaf and analyse each of your ten chosen communications against the five Cs criteria. Compile the chart under three headings, oral, written, visual. An example is given on page 81.

Use the results of your analysis of other's communications to learn from their mistakes and so improve your own communication skills. Produce an action plan listing ten positive steps that you are going to take to improve your communication skills.

WRITTEN COMMUNICATION

Details of communication received	Clear	Complete	Concise	Courteous	Correct
fax from ABC Ltd requesting confirmation of COSSH requirements for handling one of our products	✓ easy to read	✓ contained all the necessary information	✗ a bit rambling, difficult to extract relevant points	✗ no pleasantries at all	✓

7 Oral communication

Learning objectives

On completion of this chapter, you will be able to:

- listen effectively
- conduct interviews
- plan for and chair meetings
- use appropriate telephone skills
- give a presentation.

Introduction

Both listening and speaking are skills of oral communication. These activities take up about 60 per cent of the first line manager's average day. If you can improve your oral communication skills your job performance is bound to improve.

Effective listening

Hearing and listening are not the same thing. Listening is an activity; you have to concentrate to listen effectively. Listening is something that you do, not something that happens to you. We spend so much of our time listening it follows that if you can improve your listening skills you will improve your communication skills.

Why we are such bad listeners

It is not easy to be a good listener, it requires concentration and effort on the part of the listener and we often do not put as much effort into listening as we should. It is easy for your concentration to wander while you are listening, it is also easy to be distracted. We are all used to 'easy listening', for example, to the television and the radio. Usually when we are listening and relaxing we do not listen properly at all. Another problem is that we often think that we know what the person who is speaking to us is going to say and so we hear what we want or expect to hear. We have to take care to listen and not just wait for the gaps so we can have our say.

How to be a better listener

Follow the guidelines below to become a better listener.

- *Stop talking* - it is impossible to listen and talk
- *look and sound interested* - maintain good eye contact, lean forward, adopt an open posture, face the speaker, be relaxed
- *ensure that the full story is told* - probe gently with open questions (see later in this chapter for a description of open questions)
- *wait for the complete message* - listen for the central theme of what is being said, wait until the speaker has finished speaking
- *ensure that you understand* - if there is anything that you do not understand, ask questions
- *keep an open mind* - do not let your personal opinions or prejudices influence your appraisal of the situation
- *be patient* - do not hurry the speaker, show understanding.

Good listeners do not:

- confuse the speaker with what they are saying
- monopolise the conversation
- let their eyes wander from the speaker's face
- interrupt
- try to top the other person's story or joke
- criticize
- argue
- let themselves be distracted
- get emotionally involved
- speak too soon.

Activity 29

Are you a good or a bad listener? Jot down five ways in which you need to change your behaviour to become a better listener.

Speaking

Reflect for a moment on how much time you spend each day speaking: giving information, attending meetings, talking to colleagues, interviewing, talking on the telephone. You need to speak clearly, keep your message simple and clear and present your thoughts logically, all of which makes it easy for people to listen to you. It is possible to improve your speaking skills. Here are some guidelines to help you improve your speaking expertise.

You need to be clear about the purpose of the message and bear in mind who your listener, or listeners, are going to be. It is vital to use language that your listener will understand. Do not use technical jargon if your listener will not understand what you are saying. You need to be careful that your listeners will not suffer from information overload; this can happen if you give the listener too much information too quickly. Make sure that you deliver your message at a speed that the listener can cope with, concentrate on the key points, summarize to reinforce the message.

Obviously, you will adjust the way in which you speak depending on who you are speaking to and the circumstances in which you are speaking. For example, you speak in a different way when you are talking to a friend at lunch time than when you are giving a member of your team an instruction or speaking at a meeting. When talking to your friend, you will talk in a casual, informal way and let your feelings show. When speaking in more formal circumstances you will be more restrained and not reveal the way you feel as readily as when talking to a friend.

Asking questions

One of the main methods you will use to gather information is asking questions. You ask questions of your workteam, to try and find out what has happened, at interviews, of other team leaders, of your manager. There are different types of questions and different circumstances in which they should be used. Table 7.1 illustrates the different types of questions that there are and how they can be used.

Table 7.1 Types of questions

Type of question	Example	Effects
Open	Why would you like this job?	Cannot be answered 'yes' or 'no'. This encourages the respondent to provide full information. These questions are usually prefixed with why, what, where, when or how
Closed	Do you want to work overtime this Saturday?	Will be answered 'yes' or 'no'
Specific	Exactly when did you leave Lewis Limited?	The answers provide specific, detailed information
Leading	So, you really would like this job would you?	The answer expected or required is indicated in the question
Hypothetical	I always prefer to enter my production figures at the end of each shift, don't you?	These questions test reactions and depth of thought
Reflective	If June is promoted, how will you manage without one of your key operators?	These keep the person talking and elicit more information
Prompting	So, what did you do next?	This type of question helps the speaker to give you more information
Probing	What exactly did Joanne say to you?	Use these types of questions when you want to know more about a specific incident or subject

Body language

Body language is all non-verbal communication of attitude and mood conveyed by the body. The saying 'actions speak louder than words' can be very true when dealing with people face to face. When you speak to people it helps to be aware of your own body language and the body language of the person to whom you are speaking. Being aware of the meaning of other people's body language will help you interpret fairly accurately how people are feeling; this will enable you to reassure them when appropriate and provide feedback when necessary.

Beware though, of reading too much into body language and try not to interpret one act in isolation. A person might fold their arms, for example, because they are feeling cold, not because they are feeling defensive. Remember also that different cultures may interpret gestures in different ways and so it is possible to that you could be reading a message quite differently from how it was meant. It is wise to use non-verbal communication as merely an indicator of how a person might be feeling.

Facial expression

When talking to people it helps to look friendly and interested by smiling and looking welcoming, rather than looking bored or distracted.

Posture

Your posture will give an indication of how you are feeling. For example, if your shoulders are raised and your head lowered you will probably be feeling tense and negative. When your head is raised it indicates openness and interest. When your head is tilted sideways a little it indicates interest and curiosity. Arms tightly crossed show that the listener is feeling defensive. When listening, aim to sit in an open relaxed way - it will put the speaker at ease.

Gestures and movements

Try to use gestures that indicate that you are listening, such as nodding; this shows that you are listening or agreeing and encourages the speaker to continue talking. A shrug indicates indifference. If you walk briskly into a room you look efficient and determined, whereas if you shuffle into the room you can look lazy and/or depressed. Try to keep still and not fidget or you will look nervous or impatient.

Eye contact

This is one of the most powerful forms of non-verbal communication. By using eye contact you can show somebody that they have your absolute and undivided attention.

Interviewing

We looked at interviewing as a means of gathering data in Chapter 3. Here we will examine interviewing in more detail. An interview is a meeting in which the discussions that take place have an explicit objective, one party being responsible for achieving this objective. The interviewer is responsible for controlling the conversation and achieving the purpose of the interview.

There are two types of interviews:

- *Formal* - these are planned, have very specific objectives and are held in a private setting. For example, personnel interviews for selection, appraisal, disciplinary, grievance, exit and counselling purposes.
- *Informal* - these require very little specific planning or preparation and are more frequent than formal interviews. For example, those that occur when a customer calls at the enquiry desk or when colleagues ask for snippets of advice or information in the course of their work.

The five stages of an interview

All formal interviews involve five stages:

- planning
- opening
- main part
- closing
- follow up action.

Now we will examine each one of these stages in more detail.

Planning the interview

This first stage is very important as it provides the foundation of a good interview. You need to prepare yourself for the interview in three ways.

1 *Mentally* - remind yourself what the objective of the interview is. How long should the interview last? Have you got any relevant information about the interviewee? All of these elements will influence how you approach the

interview. For example, if you have very little information about the person you are interviewing, you may need to spend the early part of the interview getting information about the interviewee.

2 *Environmentally* - decide where the interview will take place and how the seating should be arranged, then check that there will be no distractions such as sun in the eyes, a wobbly chair, noise and interruptions.
3 *Materially* - gather the information that you need for the interview, do your research, making a list of points that should be covered.

Opening the interview

First impressions are crucial, so it is important to open the interview in the correct way in order to create the right kind of impression immediately. Obviously the exact type of approach will depend on the type of interview.

The main body of the interview

It is at this stage of the interview that the information you require for the outcome of the interview to be successful is extracted from the interviewee. This is achieved by asking open questions, which encourage the interviewee to talk. Pause regularly, this will also encourage the interviewee to talk, take notes if necessary. It is important that you maintain control of the interview. If the interviewee side-tracks, bring them back to the point by asking a relevant question. Regularly summarize the conversation to check that you and the interviewee understand each other. Listen to what is said, also listen for what is not said. Use appropriate body language, which indicates that you are both listening and interested. Keep an eye on the time. If time is running out ask closed questions, this will enable you to get information quickly. Throughout the interview show an interest in the interviewee and try to develop an empathy with them.

Closing the interview

Summarize what has been discussed or decided. This provides a positive finish. Make sure that you are both clear what future actions you both need to carry out. Once the interviewee has gone, take a few minutes to reflect on how effective your chosen techniques were and on ways in which you can improve your interviewing technique.

Following up after the interview

Ensure that the necessary paperwork is completed and that any follow-up work is done as soon as possible.

Good interviewers

During an interview the most important person is the interviewee and you must make the interviewee feel that that is so.

Remember the following tips to help you to be a good interviewer:

- welcome the interviewee, smile
- keep the interview objectives in the front of your mind
- use plain language, avoid jargon
- use appropriate body language – adopt an open posture, do not fidget or show signs of being bored
- establish good eye contact
- sit at 90° to the interviewee and not too far away, do not sit on the other side of a desk
- sit at the same level as the interviewee
- relax, listen, observe and interpret non-verbal signals
- seek information by asking open questions
- analyse replies
- do not talk too much
- stay calm, retain your self-control
- give your full attention to the interviewee
- follow up efficiently.

Think back to an interview that you have experienced as the interviewee. Reflect on the skill of your interviewer. Was the interview well conducted? Could the interviewer have improved their interview technique? What can you learn from the interviewer's mistakes?

See Feedback section for answer to this activity.

Meetings

We all spend some of our time at work in meetings. Meetings are important because they enable communication within the organization and with people outside the organization. The problem is that we all feel some of the meetings we attend are a waste of time. We usually feel like this after we have attended a meeting that has been badly managed. People resent attending a meeting that they feel is a waste of time; we are all too busy at work to waste such a precious commodity. Remember, only hold a meeting if you have to. If there is a more appropriate, less time-consuming way to achieve what you need to, such as sending an E-mail or a memo, then do not hold a meeting.

In this section we will look at the process of meetings to help you both contribute to a meeting and run a meeting in an effective manner.

A meeting is when two or more people get together for a specific purpose.

Activity 31

List three reasons why meetings are held at work.

See Feedback section for answer to this activity.

You might attend a meeting as a member of the group or you might be leading the meeting. You might be invited because of your job, because you have particular knowledge that is required or because you are representing a group of people.

It is likely that you will attend meetings at which you are the most senior person, meetings with other team leaders, meetings with staff who are senior to you, meetings with customers and meetings with staff from across the organization.

Obviously the way in which you behave at the meeting will depend on what type of meeting it is. However, you should always be polite, respect others and listen carefully.

Activity 32

There are advantages to having meetings. List three ways in which meetings are useful.

See Feedback section for answer to this activity.

Types of meetings

There are two types of meetings:

- *Formal meetings* – these meetings have written rules and procedures. Formal in this context means formal rules and procedures. Formal meetings have elected officials such as chairperson, secretary and treasurer. It is usual to make any comments through the 'Chair'.
- *Informal meetings* – it is likely that you will spend more of your time at informal meetings. Informal meetings will also have a chair or leader who runs the meeting but the procedures are less formal.

Procedures at formal meetings

Although, in this book, we will concentrate on looking at informal meetings, in this section we will look briefly at the procedures at more formal meetings.

Meetings are normally called by the Chair who will send out a *calling notice*, which is usually accompanied by an *agenda*. There are usually some items that appear on every agenda such as apologies for absence, approval of the minutes of the last meeting, matters arising from the minutes of the last meeting; these are usually at the beginning. These are followed by the other items which make up the agenda. The agenda usually closes with the two items 'any other business' and the 'date, time and location of the next meeting'.

For some formal meetings there needs to be a *quorum* in attendance at the meeting. This means that unless a certain number of members are present at the meeting (the number will be specified in the rules) business cannot be transacted.

In some meetings discussion cannot commence until a *proposer moves a motion*. The motion is supported by a *seconder* and, if it is approved, the motion becomes a *resolution*. If the committee wishes to alter a resolution an *amendment* is proposed; that has to be seconded and then agreed by the committee if the motion is to be altered.

If any committee member has anything to say their comments must be addressed through the Chair. Members do not talk directly to each other, but through the chairperson.

Planning meetings

If you are organizing a meeting at work it will usually be an informal meeting. Make sure you are clear on the reason why the meeting is being held. What are the objectives? Who needs to attend? How far in advance will the attendees need to be notified?

Before the meeting, however informal, let people have written notice of the date, the start time and finish time, the place and a list of those invited to the meeting. This is usually notified at the same time as the agenda. See Figure 7.1 for a sample agenda. The agenda lists the topics that are to be discussed and it should be distributed well before the meeting.

Memo

To All members of Team A
From Fred Jarvis, Team A Team leader
Date 3 September 19XX
Subject Team meeting

The next meeting of Team A will take place on Monday 17 September 19XX at 10.30 am–11.30 am in room 37b. Please will you all make every effort to attend.

AGENDA

1 Notes of the last meeting on 14 August 19XX
2 Matters arising
3 Timekeeping
4 Introduction of new shift patterns
5 Goods Inwards – new method of working
6 Any other business.

Figure 7.1
A sample meeting agenda

Chairing meetings

If you want to ensure that the meetings that you chair are successful, follow the guidelines below.

- If you are chairing the meeting you will not be able to write notes at the same time, so get someone else to do it for you
- plan well
- start on time

- create the right atmosphere
- encourage participation, be fair, give everyone a chance to speak
- do not let one or two people dominate the discussion
- listen carefully
- summarize often to check understanding
- treat everybody with respect
- be firm, maintain discipline whilst being patient and good tempered
- keep an eye on the time - do not overrun
- ensure people stick to the point.

After the meeting

You need to check whether or not you have achieved the objectives of the meeting. Ensure that the notes are written; notes of a meeting are often called minutes and are the written record of what took place at the meeting. Ensure that they are distributed to everyone that attended the meeting and to anyone else who might need a copy, for example your line manager.

Ensure that the minutes are written within 24 hours of the meeting, otherwise you (or whoever is writing the minutes) will forget what has been said. Check that all actions have been allocated to an individual or task group to carry out.

Figure 7.2 is an extract from the minutes of the meeting for which the agenda is shown in Figure 7.1 on page 92.

Obtain documentation relating to a meeting held at your workplace. Is the meeting documentation of a high quality? Does the agenda give all the essential information? Do the minutes make it clear what actions have to be taken and by whom?

Communicating on the telephone

Obviously, when you are talking to someone on the telephone, it is not possible for you and the listener to see each other. You cannot see the body language of the person you are speaking to, you do not know if they are on their own, what is happening around them; if you have never

Minutes of Team A's Team meeting held on Monday 17 September 19XX at 10.30 am–11.30 am in room 37b.

In attendance: S Jones
B Speet
F Carruthers
R Nair
C Kazmarcyk
D Smith

Apologies: G Stuart

Distribution: Team A
H Little – Factory Manager

1 **The notes of the last meeting** on 14 August 19XX were agreed as an accurate record.
2 **Matters arising** – The new telephones had now been installed and so staff were finding it easier to answer the phone. The technical support team have undertaken customer service training as suggested and an improvement in the service had been noted.
3 **Timekeeping** – Fred pointed out that some team members were getting lax about their timekeeping; late arrival in the morning and extended breaks were having a detrimental effect on production. The team agreed to improve in these areas. **Action – whole team to improve timekeeping, to be reviewed at the next meeting**
4 **Introduction of new shift patterns** – See attached sheet for details. The new shift patterns would start at the beginning of next month. We will review the success of the new system at the next team meeting. If anyone has any problems before then please see Fred.
5 **Goods inwards** – New method of working – see attached sheet for details. George and Susan explained the new system. There were a couple of suggestions for improvements that could be incorporated into the new system. **Action – Brian and Colin to meet with George and Susan so that their proposals can be incorporated before the system is introduced. New guidelines to be ready by 1 November 19XX.**
6 **Any other business** – Brian expressed concern about the regularity with which the mainframe computer system was crashing. This was becoming more frequent. **Action – Brian will keep a log over the next month of the times and duration that the system is not working. Hopefully, this will provide Fred with the evidence he needs to take up the matter with Head Office.**

Minutes prepared by Brian Speet.

Figure 7.2 Sample minutes

met face to face you will not even know what the person looks like.

You will need to speak more clearly when you are speaking on the telephone. It is very important to check that the person you are speaking to understands the message that you are conveying.

Before you make a call, always make sure that you have the information to hand that you think you will need.

When you answer the phone speak clearly, something like, 'Good morning. Black Brothers. Sally speaking. How can I help you?' This gives all the information that the caller needs. The caller has been welcomed, knows that they are talking to the right organization, knows who you are and that you want to help. Always have a pen and notepad ready so that you can take down a message, you are much less likely to forget if you write it down.

Write down what you will say when answering the telephone at work. Record yourself and play back the recording. How do you sound? Is your voice clear, do you sound friendly? Do you need to improve your telephone answering technique?

Giving a presentation

For most of us, circumstances at work mean that there are occasions when we have to give a presentation. It is also a fact that most of us hate speaking in public. As with most things it is possible to learn how to speak in public. There will be a range of occasions at which you may have to speak, for example, a retirement speech for a member of your workteam or to brief a group on a project you have been working on. Public speaking makes most of us nervous; if you are asked to give a talk or a speech, it will be important to you and the audience and you feel responsible for giving a good talk. Try not to let your nervousness show, do not fiddle with your jewellery, your papers or your tie, or play with the change in your pocket.

If you are giving a formal presentation, it will help if you consider the following points. It is vital that you prepare well.

Clear aims

Ensure that you are clear about the purpose of your talk. Think about the needs of your listeners and the context in which you will be speaking.

Content

Decide the key points you need to put across in order to achieve your objectives.

Structure

Work out a sensible and logical structure for the points you want to make. There must be:

- A *beginning* - an introduction. Make this a summary of what you are going to cover. State who you are and the purpose of your talk.
- A *middle* - put across the key points you want to make.
- An *end* - the conclusion, rounding off your talk. Sum up the main points to be remembered. Try to express them in a different way to avoid repetition. Always finish on a positive note. People will remember the last thing you say. Make sure that your talk has a definite end. Do not let it peter out.

Initially collect more material than you will actually have time to present, then choose your best material. Sort your material into information the audience must be given, information they should be given and information they could be given. Remember always take into account the background and existing knowledge of your audience.

Visual aids

If you are doing a formal presentation, visual aids can be very useful to back up what you say. They can help to get a difficult idea across and add impact and interest. You can use such equipment as an overhead projector, flip chart, white board, slide projector or a presentation package on your computer software. Remember not to talk to the visual aid you are displaying. Talk to the audience. Cover up/remove visuals that you have finished with. Use colour and shape for impact. Visuals need to be big, bold and clear. Do use pictures and diagrams to emphasize a point.

Notes

Make a brief outline of what you are going to say. Do not read your notes word for word. Keep your notes brief. It is better if they are written on postcards; number the postcards in sequence and write key words and phrases on the cards. Write statistics/quotations in full and make a note of the points at which you will show visual aids.

Prepare materials

Prepare any materials such as handouts, specimen forms and brochures in advance.

Rehearsal

Practise your delivery, check your timing and think about what words to use. Speak to the whole audience individually. Use eye contact. Practise speaking clearly and carefully. Do not forget to show enthusiasm for your subject and to smile from time to time. Dress appropriately. This will make you feel more confident.

Feedback

Encourage the audience to ask questions, but if you prefer, you can ask them at the beginning to save their questions until the end. You might find, if you are not an experienced presenter, that questions at the end are easier to cope with.

Summary

In this chapter we have covered the aspects of oral communication that impact on your daily life as a team leader. Remember that 60 per cent of your average day is spent involved in speaking and listening. Your communication skills need to be really good if you are to be good at your job. The most neglected skill is listening. It is easy to improve your listening skills. It will take longer to understand body language, but if you start to observe now you will notice how it improves your understanding of people when you are communicating with them. We have looked at interviews generally; the specific types of interview mentioned in this chapter (selection, disciplinary, grievance, exit, appraisal and counselling) are covered in the book in this series entitled *People and Self Management*. If you are reading this book to support a course of study you are undertaking in supervisory management you will no doubt be asked to give a number of presentations throughout your course. This will help you to practise your presentation skills. If you simply want to improve your presentation skills we have covered many points which will be of use to you.

Review and discussion questions

1. What is the difference between hearing and listening?
2. List five ways in which to become a better listener.
3. What are the differences between an 'open' and a 'closed' question?
4. Differentiate between a formal and informal interview.
5. List the five stages of a formal interview.
6. Describe the following terms in relation to meetings: agenda, quorum, motion, minutes.
7. How should you answer an incoming telephone call?
8. Why is it a good idea to use visual aids when giving a presentation?
9. How should you prepare the notes you will use to support a presentation you are giving?
10. 'All the meetings at my workplace are necessary, well managed and help the organization achieve its overall objectives.' Discuss this statement in the context of your organization.

Case study

You are a team leader and Martin is a member of your workteam. He has recently started to chair the safety committee meeting. You do not usually attend this meeting, but you were asked to attend when the group last met as they needed some clarification from you in relation to a technical issue.

Martin was late arriving for the meeting, when he arrived there were not enough chairs and the room was freezing cold, it took him five minutes to find another two chairs and track down a fan heater. Although Martin had issued an agenda the day before the meeting, not everybody had received their copy, apparently there had been a problem with the internal mail. Stella and David seemed to dominate the meeting, the other members of the group could not get a word in; Lucy, Emily and Christopher did not say a word for the entire meeting. You happened to know that Christopher was an expert on one of the matters that was discussed; he did try to say something at one stage but Stella started talking again and he gave up.

The meeting ran over time by half an hour. Martin had to close the meeting rapidly and two items on the agenda had to be carried forward to the next meeting. You heard Lucy and Emily complaining as they left that the whole thing had been a waste of time and nobody wanted to listen to what they had to say.

Obviously Martin needs some guidance on how to chair a meeting. You decide to prepare a fact sheet entitled 'How to Chair a Meeting', which will help Martin. Prepare the factsheet.

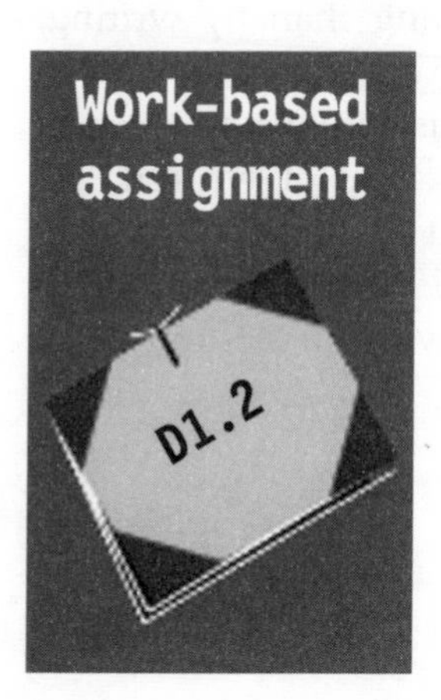

Prepare a 10 minute presentation entitled 'Communication – an essential skill for every successful team leader'.

Make sure that you plan your presentation carefully. Prepare your notes on postcards and use at least two visual aids to support your presentation.

8 Written communication

Learning objectives

On completion of this chapter, you will be able to:

- identify when it is appropriate to communicate in writing
- use a simple and direct writing style
- select the appropriate style of writing for the type of communication
- select the appropriate language to enable the reader to understand your message
- organize your writing so that your messages are easily understood
- write letters, memos and reports.

Introduction

In the last chapter we looked at the importance of spoken communication. Another way in which we communicate is in writing. Often there are circumstances at work when you have to use written communication. Many team leaders feel more confident communicating by speaking than by writing. However much you might dislike using written communication, there are some occasions when you need to write a letter, memo, report or other type of written message. This chapter has been designed to help you decide when to communicate in writing and to write more effectively when you do have to use a written method of communication.

When it is appropriate to communicate in writing

Activity 34

Look at the following messages and decide whether it would be appropriate to communicate by speaking, writing or both.

The message to be communicated	Speaking and/or writing
Noxious fumes have escaped into the factory, it must be evacuated immediately	
A member of your workteam is carrying out a task without proper regard for health and safety	
You are walking around the factory and notice that a work area is very untidy	
You witness an accident in the factory	
You need to inform 100 of your customers about a changed specification for one of your products	

See Feedback section for answer to this activity.

You can see that there are some occasions when you need to communicate by speaking and many when you need to communicate in writing.

It is quite often necessary to back up spoken communication with written communication. For example, you may need to speak to somebody about not paying proper attention to health and safety, but if it is the second or third time you have noticed them contravening the regulations, you might write to them formally, to emphasize how serious you consider the matter to be. If you need to take disciplinary proceedings at a later date, the evidence of the written communication may be helpful.

The advantages of written communication

We looked at the advantages of written communication in Chapter 6. Four main advantages were identified. Can you remember the advantages? List them below.

See Feedback section for answer to this activity.

You will recall that there are several disadvantages to written communication. These are:

- written communications take time to produce
- they tend to be formal and distant
- they may be misunderstood
- it is not possible to get instant feedback
- once you have committed yourself to writing, it can be difficult to change the message
- written communication does not allow for a rapid exchange of opinions, views or attitudes.

There are a number of situations when you should write rather than speak. You need to make a judgement about when it might be better to write something down rather than speaking to a person face to face or on the telephone. Although writing takes longer than speaking, there are circumstances when, in the long run, it will cost you much less time by using a written form of communication now than try to rely on memory later. Or having to explain everything more than once because the spoken message is not clear.

Sometimes it is essential to keep a written record for legal reasons, such as recording accidents in the accident book. Sometimes it is a good idea to put something in writing to cover your own back. For example, in a scenario such as the one described above, let us say that one of your workteam is constantly removing his protective goggles when you are not around. You have spoken to him on three occasions and written to him twice. One day he has an accident which results in him losing an eye. You can point out that you have spoken to him and written to him about the importance of wearing appropriate personal protection. It would be difficult to blame you for what happened, as you have evidence of how hard you tried to make your team member wear his goggles.

Are there any procedures that you or your team have to follow at work that should be in writing, but are not? If there are, why not either write the procedure yourself or make arrangements for the appropriate person to write the procedure.

Clear and simple communication

In Chapter 6 we looked at the five 'C's of communication.

Activity 36

List the five 'C's of communication.

See Feedback section for answer to this activity.

Now we will look at each one of these in more detail.

Clear

You must aim to use clear, simple, straightforward language that your reader will understand. Do not feel, just because you are writing, that you cannot use ordinary everyday words. Simple language is much easier for the reader to understand and gives a better impression.

Complete

Make sure that you say everything that you need to say. Ensure that you read through your message when you have finished it, to be certain that you have not forgotten anything.

Concise

Do not say more than you need to. Notice how much shorter the letter suggested in the Feedback section is than the letter on page 104. You can see that clear, simple writing is much more concise. Keep your sentences short and to the point. Use a logical structure: introduction, the body of the message in the middle and a polite end.

Activity 37

Rewrite the following letter using simple clear language.

Dear Mr James

With reference to your recent letter dated 17 February 19XX.

At this present moment in time, we are experiencing major difficulties in relation to deliveries from one of our key suppliers, who provides product 234/2367/897/458/3456. This is an essential component for the product 2679/78/UP, which we supply to you.

I have to say that this difficulty is unprecedented. Consequently we have not yet calculated how we can deal with this fundamental difficulty. In the near future, (I have a meeting with our suppliers this week), I will be able to ascertain how long the inventory supply problem will last and so be able to communicate new delivery dates to you.

Please accept my sincere apologies for the significant inconvenience that has been caused to you. I will communicate with you by telephone as soon as I can expedite matters, which will be towards the end of this week.

Yours sincerely

David Pratt

David Pratt
Production Supervisor

See Feedback section for answer to this activity.

Correct

You need to make sure that your facts are correct when you are going into print. You also need to be sure that your grammar and spelling are correct. This is not too difficult these days if you are using a word processing package as most of these have spelling and grammar checks. If you need to write a note or a memo by hand make sure that you have a dictionary handy.

Courteous

Always be polite. Remember to thank your customers for their business. Remember your organization needs to retain all its customers.

Letters

You may have to write letters all the time at work or it may be something that you only need to do occasionally. If you do not have to write letters very often, it can be quite difficult to write them. The guidelines below provide some practical advice on how to construct a good letter.

- Decide on the purpose of the letter.
- Collect all the relevant information together.
- Plan the letter so that it has a logical order, letters naturally fall into three sections:
  1. the first paragraph which states the purpose of the letter
  2. the middle paragraphs which develop the purpose of the letter
  3. the final paragraph which indicates what response or action is required and concludes the letter.
- If you are answering a letter, refer to it in your reply. For example, you might write, 'Thank you for your letter dated 17 November . . .'. If you are answering a letter which contains a reference, make sure that you include the reference in your reply.
- Keep the letter as brief as possible.
- Be polite.
- Use simple, straightforward English, no jargon.
- Consider who you are writing to and use appropriate language.
- If you know the name of the person you are writing to, use it.
- Give your letter a title if that will make the message easier to understand.
- You can use subheadings and/or numbered paragraphs in a letter, if it will make the message easier to understand.

- If you begin Dear Sir or Dear Madam finish 'Yours faithfully'. If you address the letter to a named person, for example, 'Dear Ms Patel' or 'Dear Mr Froggit', finish 'Yours sincerely'.
- Check the accuracy of the letter.

There are many ways to set out a business letter but Figure 8.1 is an example of a 'fully blocked' layout which is popular, modern and easy to learn.

Memos

Memos ('memo' is a shortened version of 'memorandum') are used for internal communication within the organization. You never send a memo to someone outside the organization. A memo, whatever the length, always has the same layout.

Some organizations have pre-printed memo forms and some use a standard memo format from the word processing package used by the organization. Memos can be sent to one or more people. Usually one memo only deals with one subject. As with letters, keep your language simple and use subheadings and numbered points if it will help you make your message clearer to the recipient. Figure 8.2 gives an example of a memo.

Reports

You may need to produce reports on a regular basis at work, or it may be something that you do rarely or not at all. Whatever your level of experience, there will be some information that will be useful to you in this section. Report writing is probably the most difficult type of written communication that you may have to produce. Follow the guidelines below to help you produce a high quality report.

- Ensure that you are clear about the reason why you are writing the report and what you want to happen when your report has been read by the recipients. Reports are often required to provide information or investigate a problem.
- Ensure that the style you use for the report is appropriate for the audience.
- Always check to see if a similar report has been produced before. This will provide you with guidelines on any standard formats or house style that is generally used in your organization.

ABC COMPANY LIMITED
Fulshaw Industrial Estate, Fulshaw
Wolverhampton, WV99 ST5
Tel: 01445 12345
Fax: 01445 67890
E-mail: abc.@virgin.net

Letterhead

Our ref: 00479 — Sender's reference

Your ref: JK/AD — Recipient's reference

19 October 19XX — Date

Mr J Kershaw
Walnut House
14 Marple Road
Nantwich
Cheshire
CW25 6ZZ

Name and address of recipient

Dear Mr Kershaw — Salutation

Alpha Floor Covering — Subject heading

Main body of the letter

Thank you for your recent letter enquiring about our new Alpha floor covering. I am delighted to enclose a copy of our most recent brochure, which illustrates our new range of designs.

We keep most lines in stock and so when you have made a decision about which floor covering you prefer we can arrange to lay it very quickly, at a time which is convenient to you. Many of our customers prefer us to lay the floor covering at the weekend so that they can minimise disruption to production. I will be happy to arrange this for you.

If you have any questions or require any further assistance, then please get in touch with me. I will be happy to help in any way I can.

I look forward to receiving your order.

Yours sincerely — Complimentary close

Denise Smythe — Signature

Denise Smythe — Printed name
Customer Liaison Officer — Designation

Enc: Alpha Brochure

Figure 8.1 A sample letter

Memorandum

To: Team leaders
From: David Jones
Date: 1 March 1998
Subject: Stationery supplies

I am introducing a new system for the issue of stationery. From now all requests for stationery will have to be given to Jim Flower on a stationery requisition form. A supply of forms is attached for your use.

Jim will deliver the stationery that you have requested to you on the day following the receipt of the requisition.

If you have any queries about the new system, please have a word with Jim.

Thanks

Enc. Ten stationery requisition forms

Figure 8.2
A sample memo

Investigate 17

Are there any standard report formats in use within your organization? Get hold of a couple of reports from work. Compare them against the guidelines given in this chapter.

How to start

You will probably need to do some investigative work and make notes as you go along. You will have to get the information that you need that relates to the issue you are writing about. The facts that you collect will need analysing. You may wish to use graphs and charts to illustrate some of the information you are using (see Chapter 10). If you do use methods such as graphs, tables and charts to show information, place these in the appendices and refer to the relevant appendix in the report. You will use the information you have collected and analysed to write the investigation and analysis section of your report. Decide on the central and secondary issues. Gather all the facts. Collect and organize the material you will be using.

The report structure

The finished report will need to be presented in a logical format. The actual format that you will use will depend on the length of the report and whether or not there is a house style. An example of an appropriate structure is given below:

- Cover
- Contents
- Introduction
- Summary
- Investigation and analysis
- Conclusions
- Recommendations
- Appendices.

Contents

Always do a contents page for your report, it might look something like the contents page shown in Figure 8.3.

CONTENTS

SECTION	TITLE	PAGE NO.
1	Introduction	1
2	Summary	2
3	Investigation and analysis	3
4	Conclusions	8
5	Recommendations	9
6	Appendices	10

Figure 8.3
Specimen contents page

Introduction

Write the introduction. In this section describe the purpose of the report and how you set about collecting the information.

Summary

A lengthy report will have a summary. This briefly says what the report is about and what the major conclusions and recommendations are. Write the summary last but place it at the beginning of the report. The idea is that the readers can quickly and easily read an overview of the content and grasp the main messages.

Investigation and analysis

This is the main body of the report. The headings you will use for this section of the report will vary depending on the type of report. You may have sections headed Investigation and analysis, but other headings may be more appropriate. Use subheadings and numbered paragraphs. This will make the report easier to read.

Conclusions

After you have analysed the information in your report, you will be ready to form some conclusions. State your conclusions clearly in this section.

Recommendations

Recommendations should be brief. Ensure that your recommendations are based on the
evidence you have given in the Investigation and analysis section of the report. Ensure that you have made clear exactly what you think should be done.

Appendices

Put any supplementary information in the appendices, such as graphs, charts, detailed calculations, diagrams. You must not clutter up the main report with this detail. Just refer to the appropriate appendix in the main report.

Final check and follow up

Check the report for mistakes and to make sure that it still meets your original purpose. It is essential that you follow up the report to try to ensure that appropriate action is taken.

Summary

We have identified some occasions when it is essential to communicate in writing, as well as identifying the advantages and disadvantages of using written communication. Many team leaders avoid using written communication and are more confident when they communicate by speaking. This chapter has provided you with some simple guidelines for writing letters, memos and reports at work. Remember to write in simple and direct language and always check your communications against the five Cs.

Review and discussion questions

1 List three advantages of written communication.
2 Why is it sometimes important to communicate in writing to 'cover your own back'?
3 List the five Cs of communication.
4 If a letter is addressed 'Dear Sir', should the letter close with 'Yours faithfully' or 'Yours sincerely'?
5 List the four pieces of information that are at the beginning of every memo.

Case study You are a team leader working for J. Smith Limited at Smith House, Smith Street, Smithfield, Staffordshire ST47 8BX. You have been asked by your manager, Sally Smart, to write to one of your suppliers, Beta Limited, The Industrial Estate, Regan Road, Solihull, SO19 7TY, to ask them why the last ten weekly deliveries that you have received from them have been delivered late. They were late by between 2 and 5 hours. On two occasions you have had to stop the production line because you did not have the raw materials you needed to continue production.

You have asked the drivers who brought in the deliveries why they were late and one of the drivers was extremely rude to you and a member of your team. At a recent production meeting you suggested finding a new supplier for the product and were given the go ahead to start making some enquiries. You have spoken to Mary Frank the Sales Liaison Officer on four occasions to complain about the late deliveries, but the situation has not improved. Sally has asked you to write to the Sales Manager at Beta, Fiona McDonald.

Draft out the letter you will write to Fiona.

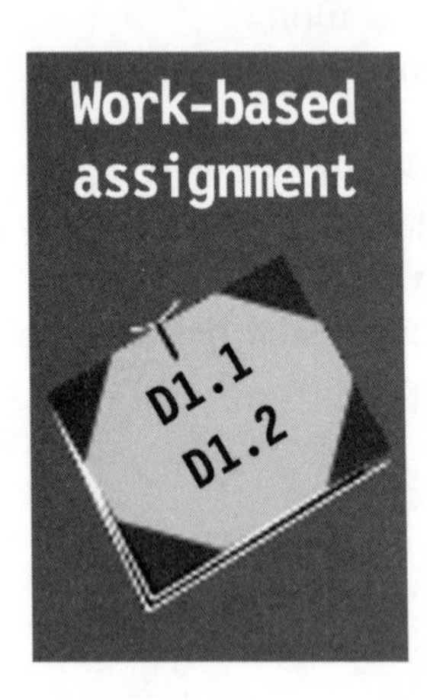

Select a problem at work that you have been meaning to investigate for some time (it should be a small problem). Investigate and analyse the problem that you have selected and write a brief report to your manager which recommends what action should be taken to deal with the issue that you have identified.

9 Communication in organizations and workteams

Learning objectives

On completion of this chapter you will be able to:

- differentiate between different types of communication within organizations
- conduct a team briefing
- give clear instructions
- give concise orders
- give appropriate feedback.

Introduction

Communication within organizations is essential if the organization is to function successfully. Messages need to be transferred from managers to the staff who work in the organization and from the staff to the managers. You need to pass on information to your workteam, to give orders, requests and instructions. You need to give your opinion and listen to the opinions of others, you need to discuss ideas about better ways of doing things. You need to motivate your team, control the workflow, deal with grievances and discipline where it relates to your team. All of this is achieved by communicating with your team and other groups within the organization such as management, other workteams, the safety committee, a project team and so on. In this chapter we will look at the way communication works within organizations; a better understanding of communication within organizations will enable you to communicate in the most effective manner in your role as a team leader.

Communication systems

Your organization will have a variety of communication systems in place. These will probably have been built up over a number of years and will be designed to ensure that messages are passed around the organization.

Think about a typical day at work. Complete the chart below by listing the type of information which is communicated to you and how it is communicated.

Type of information	How it is communicated to me
For example: Daily production figures	Computer printout delivered to me at the end of each shift

Communication within your organization will flow both vertically and laterally. Vertical communication can be divided into downward communication and upward communication.

Downward communication

This is where communication flows from one level in the organization down to a lower level. For example, information passed from your manager to you or information passed from you to your workteam is downward communication.

Upward communication

Upward communication goes up to a higher level in the organization; this could be a message from your workteam to you or from you to your manager.

Lateral communication

This is where communication takes place between people or groups at the same level within the organization. For example, communication between members of your workteam or a meeting of team leaders. Lateral communication is particularly important in coordinating the work of the organization. It ensures, for example, that your workteam is working together, that each person knows what the other person is doing. Sometimes systems for lateral communication are established by the organization but sometimes they are set up by people within the organization because the formal systems that are in place are not working very well. For example, groups of team leaders sometimes arrange to meet on a regular basis themselves to provide mutual support to each other, discuss common problems and to assist with coordinating the workflow.

Activity 39

Complete the following chart by ticking the appropriate boxes to indicate whether the communication described is 'vertical/upward', 'vertical/downward' or 'lateral'.

Details of the communication	Vertical/ upward	Vertical/ downward	Lateral
A briefing meeting for team leaders where your manager briefs you about a new customer			
A meeting of team leaders to discuss common problems			
A briefing sheet to all staff from the Managing Director			
A memo from you to your team			

See Feedback section for answer to this activity.

Communication channels

These are the channels established within an organization through which information is passed around the organization. Two types of communication channels exist within every organization: formal and informal.

Formal communication

Your organization will probably have an organization chart similar to that in Figure 9.1 which shows the formal communication systems.

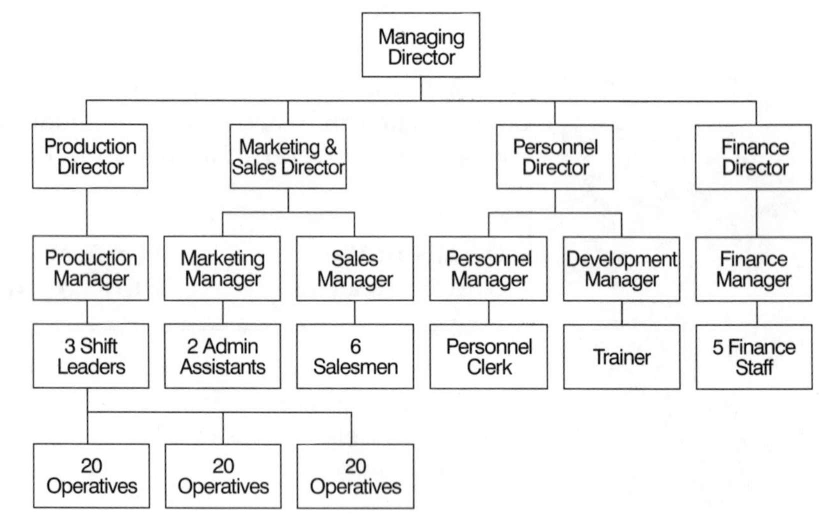

Figure 9.1
An organization chart showing formal communication channels

Informal communication

Informal communication channels are known as the grapevine. Communication via the grapevine could not be shown on any organization chart. The grapevine skips levels within the organization and meets the social needs of people who work together, for example, the group who go for a drink after work on a Friday, the darts team or the group that have coffee breaks at the same time. These types of informal communication channel are not recognized or controlled by management; they serve the needs of the people in them, not the needs of the organization. The grapevine is very powerful within organizations. Research has shown that the grapevine is about 75 per cent accurate; the grapevine thrives on rumours and gossip. As a team leader it is important that you understand the importance of the grapevine in your organization so that you can try to dispel rumours and gossip that are untrue by providing accurate information to your workteam.

Employee communication schemes

Some more forward-looking organizations have introduced formal communication schemes. Often these are not just communication schemes, but also provide for consultation. Communication is about providing and exchanging information to enable the organization to function more effectively. Consultation is about management and employees jointly examining and discussing issues that concern them both. Some organizations now have a communications and consultation policy. ACAS (Advisory, Conciliation and Arbitration Service) suggest that a good policy should set out:

- a clear statement of policy, including the purpose of communications and consultation, the fact that it is an integral part of every manager's job and the importance of communication as a two-way process and not a one-off exercise
- responsibility for communication at each level
- the methods of communication
- arrangements for consultation and participation
- arrangements for training managers and employees in the skills and processes of communications and consultation
- how the policy will be monitored.

Does your organization have a communications and consultation policy? If so, get hold of a copy and examine it. If not, how is information communicated and how are you consulted in your organization? Do the current methods work? How could communication and consultation be improved?

Team briefings

Team briefing is a method of communicating to all levels within an organization. This is done by holding a series of meetings regularly, usually each month. The meetings start at the top of the organization, say at board or senior management level, and it is at this meeting that the communication brief is produced. Team briefing enables the leader of each group within the organization to communicate information that is relevant to their teams on a regular basis. A team briefing does not usually last for more than 30 minutes.

The whole organization is carefully examined and broken down into logical briefing groups. The groups are briefed and they in turn brief other selected groups until the whole organization has been briefed.

What team briefings do

Team briefing enables managers and team leaders to inform staff of changes that are likely to affect their job. Team briefing is a method of planned organizational communication. It has been designed to improve the quality of information received by staff and to encourage, not only downwards communication, but upwards communication and participation. Staff are encouraged to ask questions at the briefing and make suggestions and give ideas; these are recorded and passed up the organization. Any questions and discussions should centre around the subject of the briefing and all briefing groups are encouraged to participate in the briefing sessions. It is very important that the sessions are properly managed and that discussions are focused on the subject of the briefing. They must not turn into moaning sessions or the value of the team briefing system will be lost.

How team briefings are structured

Dates for the monthly team briefings are usually set well in advance; it is important that team briefings take place at the arranged time and are not cancelled or rearranged. The team briefings are carried out by managers and team leaders. The size of the group to be briefed will vary, usually from 4 to 15 people. The team leaders attend a briefing given by their manager, the information received is summarized in the team leader's own words. The team leader ensures that the information in the briefing is related in a way which is relevant to the people working in their department.

At the team briefings, briefing sheets should never be handed out: the objective is to encourage the staff to participate and ask questions, this will not happen if the team leader stands there reading out an information sheet. Team briefing is a structured system of organizational communication. For it to work the briefers must be skilled in face-to-face communication techniques and all managers and team leaders must be committed to the system.

There are two parts to the team briefing. The first is known as the 'core information'. This consists of the briefing information that has come down from the top of the organization and should represent about 30 per cent of the briefing. There are many subjects that may form the basis of a team briefing, for example, financial results, organizational mission statement, time keeping, dealing with customer complaints, correcting rumours and many more. The core brief should consist of no more than four points. The second part of the briefing is 'local information' and represents the remaining 70 per cent of the briefing.

The information discussed at a team briefing is often grouped into four areas known as the four P's: progress, people, policy and points for action.

- *Progress* – People need to know how they are doing. Is their team performing better or worse than the other teams? How is the organization performing? What are their targets? And so on.
- *People* – Such things as vacancies, visitors, training courses, promotions, absenteeism levels, overtime levels and other people-related activities.
- *Policy* – Staff need to be briefed about policy changes and sometimes reminded about existing policy, for example, pension rights, health and safety, equal opportunities.
- *Points for action* – This covers the points that the briefers, team leaders in your case, want to progress and can cover anything that you want to progress in your department such as improving housekeeping, reducing wastage, improving safety. You can use team briefings as a method of promoting discussion and encouraging the involvement of your team in the issues that are affecting the performance of your team.

Classify the four examples of subjects covered at team briefings, given above, into one of the four categories of progress, people, policy and points for action.

Subject covered at team briefing	Four 'P' classification
Financial results	
Organizational mission statement	
Time keeping	
Dealing with customer complaints	

See Feedback section for answer to this activity.

Preparing for a team briefing

It is a good idea to keep a briefing folder. Keep notes relating to the last six months' briefings in the folder. Collect information about people, the progress of your department and any points for action. The information that you have collected will inform the content of the 'local' part of your brief. It is better to write the 'local' part of your brief before you attend your briefing session with your manager so the content is not affected by the 'core' brief.

When you attend your own briefing session ask as many questions as you need to. Take the opportunity to discuss matters with other team leaders. Interpret the brief that you have received from your manager into your own words, but ensure that you keep the main message. After you have written your brief, pass it on to your manger who can check it and monitor it, to ensure it is relevant and accurate.

You now need to consolidate the information you have down to five to seven points. The final brief must be in your own words and must last no more than 30 minutes, including a question and answer session.

Giving a team briefing

You need to communicate with your team, using the structured team briefing method to get across all the information that you need to and create an informal atmosphere, which will encourage the team to ask questions and participate. It is not like giving a presentation. It is more informal. The following guidelines will help you to produce a good team briefing.

- Keep your notes in front of you
- take your briefing folder to the briefing
- make sure you know if anybody is missing
- welcome the team
- make sure everybody can hear you
- try to put them at ease
- tell them to ask questions
- cover your main points - maximum seven
- ask questions of the team, to make sure that they have understood you
- end on a positive note
- ensure that the discussion and questions relate the content of the brief
- summarize at the end
- thank your team for attending
- notify the team of the date of the next brief.

Giving orders and instructions

You will have to do this regularly in your every day working life. If your team is to perform well it is important that you give work instructions as effectively as possible. Before giving an instruction, a few moments thought on the best way to word it can reduce the possibility of misunderstanding and save you time in the long run.

Communicating instructions

Of course, it is important always to be polite when giving instructions to your workteam. Most of us ask people to do things, we do not tell them. For example, you might use phrases such as:

'When you have finished what you are doing, will you . . . ?'
'I would like you to . . . ?'
'Would you . . . ?'

It is also good manners to finish your request with a thank you or please. The way in which you phrase an instruction does have an impact on how the recipient of the instruction feels.

Here is the same instruction phrased in two different ways. How does each instruction make you feel?

Instruction 1: 'Get those orders out by 12 noon at the latest, or we'll be in a right mess.'

Instruction 2: 'The orders that you are working on at the moment are very important. If they are not processed by 12 noon we will be short of raw materials next week. Please will you do your best to have them finished by then. If you run into any problems, let me know in plenty of time. I'll check back with you at 11.30 to see how you're progressing. Thanks.'

See Feedback section for answer to this activity.

You can see how important it is to phrase your instructions in the right way, so that your team feel valued and produce their best work. They will do that if instructions are properly given and they understand why tasks have to be completed. You need to take into account the following guidelines for giving good instructions: what, who, why, how, when, check understanding, offer support, follow up and thanks. Each one of these is considered in more detail below.

- *What* – You need to be clear in your own mind exactly what you want to be done. You can only give a clear instruction if you are clear about what you want doing.
- *Who* – You must be specific about who you want to be responsible for doing the task. It is no good saying 'one of you needs to . . . ', specific named people need to be responsible for the tasks that need to be carried out in the workplace.
- *Why* – If you explain to the person that you are asking to carry out the task why you want it doing, they will appreciate the need for the task to be done.

- *How* - In some circumstances it is essential to explain how you want the job doing. This is not always the case, there are many circumstances when a person knows how to do the task and there is no need to explain how it needs to be done.
- *When* - You need to make it clear when the task has to be completed by. People cannot always categorize jobs in order of importance if they do not know what the deadline for the task is.
- *Check understanding* - You must check that the instruction has been understood.
- *Offer support* - Make sure that the person carrying out the task knows where to go for help if they run into a problem. Offer support to your workteam so that they will not waste time trying to solve a problem that somebody knows the answer to.
- *Follow up* - After an appropriate time, follow up, check how work is progressing.
- *Thanks* - A job well done always deserves a word of thanks. This will make your team feel appreciated and motivate them to do a better job.

The example below looks in detail at how you can formulate an instruction.

Jim is a team leader.

- *What* - Jim needs the 'screw feed hopper' cleaning out before the next production run.
- *Who* - David and Stuart have just finished tidying up after having completed the last production run. Jim wants them to do the job.
- *Why* - The machine needs to be contamination-free to produce a high-quality product.
- *How* - Jim has a written 'instruction sheet' that describes how the task is to be carried out. The sheet is designed so that it requires a signature from the operatives cleaning the machine when each stage of the process is completed.
- *When* - The task needs to be completed before lunch.
- *Check understanding* - The sheet will be returned to Jim after the job is completed, he will check that all the tasks have a signature next to them.
- *Offer support* - Jim has cleaned the 'screw feed hopper' many times. If David and Stuart run into problems they can ask him for guidance.

- *Follow up* - Jim will check that the machine is clean and producing contamination-free product.
- *Thanks* - Following completion of the production run, Jim thanks David and Stuart for an efficient job which has been well done.

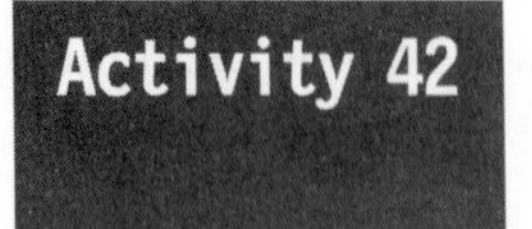

If you were Jim how would you give the instruction regarding the 'screw feed hopper' cleaning.

See Feedback section for answer to this activity.

Giving orders

Sometimes it is necessary to give orders, particularly in an emergency situation. Orders are short. The person or people that you are talking to are 'told' what to do. Examples might be, in a no smoking area with a high fire risk, 'Put that cigarette out immediately'. 'Evacuate the building now'. There is no time to be polite. The order needs to be followed immediately; usually orders are used in potentially dangerous situations.

Giving feedback

Feedback is a continuous process by which people let us know about the effects of our actions. They may do this consciously, by telling you how they feel, or unconsciously, by changing their behaviour towards you, which you may or may not notice. As a team leader it is important that you can both receive and give feedback. We will concentrate here on looking at how you can feed back to your workteam.

Most organizations have an appraisal system which enables managers and team leaders to feed back to their team members. But feedback is much more than this. It should not only happen at appraisals. It should be ongoing. Feedback is much more effective if it is given directly after an event and not stored up to be dealt with at the annual appraisal.

Positive and negative feedback

Positive feedback is easy to give. It is about congratulating somebody for a job well done. It is one of the best methods of motivating your team and so improving or maintaining high performance. The problem with positive feedback is that all too often we forget to give it.

When did you last recognize that somebody had done a good job by telling them? Do you give positive feedback as often as you could and should?

It is likely that you felt that you did not give positive feedback as often as you could. Make a mental note to give positive feedback more often in future.

Negative feedback is when you need to talk to a person, usually one of your team, about poor performance. You need the person to change their behaviour so that they will do a better job in the future. Giving feedback to your team is the best method of improving their performance, but you must be careful not to undermine and demoralize your team member by giving feedback insensitively or with insufficient thought or preparation.

Feedback and criticism

Feedback is not the same as criticism. Feedback is positive and criticism is negative. Feedback and criticism are contrasted in Table 9.1.

Table 9.1 The differences between feedback and criticism

Feedback	**Criticism**
Focuses on incident that occurred	Focuses on the person
Looks to the future and how it is possible to improve	Looks to the past
Looks for joint solutions to problems	Allocates blame
Makes specific comments	Makes generalizations
Approaches the session in a positive, friendly manner	Approaches the session in a hostile or aggressive manner

Guidelines for giving feedback

One of the problems with feedback is that we, managers, are often very uncomfortable about giving feedback.

Activity 44

List three reasons why you feel uncomfortable about giving feedback to a team member.

See Feedback section for answer to this activity.

You need to overcome these concerns about giving feedback. You will be able to if you and your team believe that feedback is a positive process. If you give feedback honestly, specifically and sensitively, feedback will improve performance.

You need to handle giving feedback very carefully. You must not be aggressive or angry. Before you give feedback you must be clear about exactly what you want to achieve. You will probably want to agree some outcomes with the person you are talking to. Be clear about these. Write them down. You must be clear about the message that you want to give to the person. You also need to decide how you want to conduct the session: you must give the person you are speaking to an opportunity to present their own viewpoint. Listen carefully and yet ensure that you achieve the outcome you need.

There are five key steps which will make giving feedback easy.

1 Maintain the other person's self esteem.
2 Feedback as soon as possible after the event.
3 Feedback in relation to the specific event.
4 Discuss why it was not good.
5 Decide together a better way of doing it.

See below for more detail on giving feedback.

- *Maintain the other person's self esteem* – Take a positive approach: the person can improve their performance in this area. Make sure the person understands why it was not good enough and the benefits of overcoming the problem. Ensure that they understand that you value them as a person. Be friendly, confident and enthusiastic.

- *Feedback as soon as possible after the event* - It is important not to delay. The sooner the better. The incident is fresh in both your minds. You will be less likely to put it off and possibly not do it at all!
- *Feedback in relation to the specific event* - Only feedback on one event. Do not take the opportunity to deal with any other issues.
- *Discuss why it was not good* - Listen to what the other person has to say. Talk about how you felt about it. Deal with what happened, not the personalities.
- *Decide together a better way of doing it* - Discuss how the person can improve. Ask them for suggestions. Be positive. Offer support.

Summary

In this chapter we have looked at the importance of communication within your organization and more specifically in your workteam. We have covered several aspects of organizational communication such as communication systems, communication channels, employee communication schemes and team briefings. In relation to your job as a team leader we have covered giving orders and instructions and giving feedback. Other aspects of communicating within your team are covered in the book in this series *People and Self Management*.

Review and discussion questions

1. Differentiate between vertical and lateral communication.
2. Explain the difference between formal and informal communication.
3. What is the difference between communication and consultation?
4. In relation to 'team briefings', what are the four 'P's?
5. Differentiate between orders and instructions.
6. Describe positive and negative feedback.
7. Contrast feedback and criticism.
8. What are the advantages of giving immediate feedback?
9. Why is it important to handle giving feedback carefully?
10. List the five key steps which you should follow when you give feedback.

Case study

You are a supervisor in an office suppliers. You witness the following conversation between Mary, a member of your team, and a customer who bought a fax machine from you last week.

Customer	'I am having difficulty understanding the instructions for this fax machine that I've just bought from you.'
Mary	'Yes, they are difficult to understand.'
Customer	'Well, er, I wonder if you could explain to me how I should set the machine up.'
Mary	'It is all in the instructions, you just have to read them carefully.'
Customer	'Yes, but it would be much quicker if you could explain what to do.'
Mary	'Well, it's not really my job to explain how to set up the machine up for you, and I'm very busy at the moment.'
Customer	'I'm busy as well. I don't think you're being very helpful.'
Mary	'Okay pass the machine over. I'll show you what to do.'

You are absolutely furious at the way Mary has dealt with this customer. You have spoken to her twice this week about her attitude to customers. You have carefully explained to her that her role is to provide the highest quality customer service. It just does not seem to sink in. You decide to feed back to Mary as soon as the customer leaves. Explain how you will give feedback to Mary.

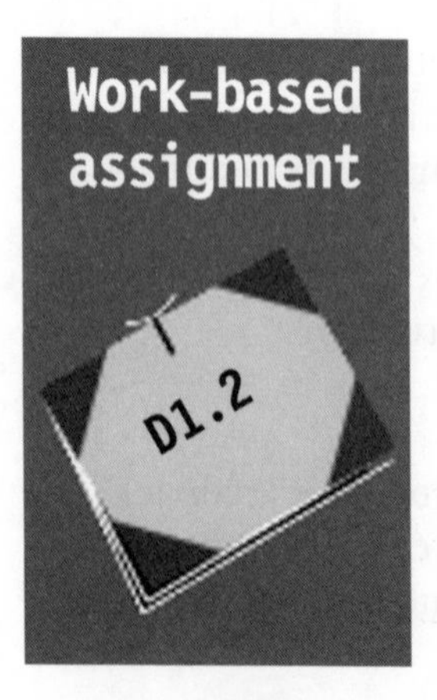

Select a written instruction at work that is not as clear as it should be. Re-write the instruction so that it is appropriate and easy for everyone to understand.

10 Analysis and interpretation of information

Learning objectives

On completion of this chapter you will be able to:

- recognize the dangers of 'information overload'
- use statistics to analyse and interpret information
- present information in diagrammatic formats
- identify trends in information.

Introduction

Whilst information is a very valuable resource, it is important that users are able to extract the relevant pieces of information from all that is available. In Chapter 5, you have learnt about the various different ways in which information can be communicated electronically. With so many opportunities now existing for access to information, it is easy to imagine that users are faced with far more information than ever before – some of it important to the decisions they are making, but much of it not.

If you are one of the very many people who finds it difficult to handle numbers with confidence, then you will appreciate how difficult it is to make decisions which involve using numeric data. The use of statistics can help to present information in different ways, so that it can be analysed and interpreted as easily as possible.

Percentages

'Raw' numbers on their own can often give a false impression, especially when they are large. Converting them to percentages is sometimes helpful.

Consider the following list of figures of the number of employees in an organization:

Department	**Number of employees**
Sales	43
Distribution	9
Canteen	6
Wages	4
Personnel	8
Research	10
Production	85
Administration	11
Security	3
Purchasing	9

A first glance at that list will give you the impression that the majority of workers are in Production. Not true. Yes, there are more workers in Production than in any other department, but there is a larger number of workers who are NOT in production. Converting the figures into percentages of the total would give a clearer picture.

The total number of employees in the workforce is 188. To calculate the percentage of the total who work in Production, you divide the number of workers in that department by the total workforce, and multiply by 100:

$$\frac{\text{Number of workers in the department}}{\text{Total workforce}} \times 100 = \frac{85}{188} \times 100$$

$$= 45\%$$

This makes it clear that the number of Production workers is well under half of the total.

Rounding and the level of precision

The exact percentage is 45.21276. It is unlikely that you will need to be this precise, so the figure can be 'rounded' to the nearest whole number, which is 45. If the answer had come out at 45.5 per cent, or higher, it is common to round the number up to 46. If it came out at 45.4999, it is common to round it down to 45.

Sometimes it is not appropriate to 'round' to a whole number, if the fraction is important. For example, if you are awarded a pay rise of 2.4 per cent, you would not be pleased if it were rounded down to 2 per cent as you would get less in your pay packet. If you are awarded a rise of 2.6 per cent, it is unlikely that your employer would round it up to 3 per cent, as the extra cost over a large number of employees would be too high. So the percentage would need to be more precise.

Calculate the percentages of the total workforce for each of the other departments in the same way as for the Production department. Remember the formula:

$$\frac{\text{Number of workers in the department}}{\text{Total workforce}} \times 100$$

See Feedback section for answer to this activity.

If you add up all the percentages, including the Production department, it should come to 100 per cent. In this example, it does. But where you round numbers up or down, it is possible that the total will only come to 98 per cent or 99 per cent. In such cases, it is usually acceptable to add an extra percentage point to the largest department, as it will make no difference to the overall picture.

Using percentages to calculate a number

Percentages can also be used the other way round, to calculate a number from another number. For example, using the organization above, suppose you are told that 25 per cent of Production workers were female. How many is that? To find the answer you mutiply the total number of production workers by the percentage, and divide by 100:

$$\text{Total number of production workers} \times \frac{25}{100} = 85 \times \frac{25}{100}$$

$$= 21 \text{ females}$$

Again, you might have to 'round' the result to the nearest whole number. The exact answer to the above calculation is 21.25 - but you cannot have 0.25 of a person!

Work out the number of females in the other departments, given the following percentages:

Sales (55 per cent), Distribution (20 per cent), Canteen (67 per cent), Wages (25 per cent), Personnel (50 per cent), Research (80 per cent), Administration (75 per cent), Security (33 per cent), Purchasing (45 per cent).

See Feedback section for answer to this activity.

Tables

Lists and strings of numbers are often difficult to absorb. It is sometimes more useful to show the information in a table. The information you have been given, or have calculated, above, could be shown as in Table 10.1.

Table 10.1

Department	All employees		Female employees		Male employees	
	Number	%	Number	%	Number	%
Sales	43	23	24	55	19	45
Distribution	9	5	2	20	7	80
Canteen	6	2	4	67		
Wages	4	3	1	25		
Personnel	8	4	4	50		
Research	10	5	8	80		
Production	85	45	21	25		
Administration	11	6	9	75		
Security	3	2	1	33		
Purchasing	9	5	4	45		
Total	188	100	78	41		

The table shows the breakdown between departments, and also between females and males in each department. Note the final figure in the column showing the percentage of females in each department; this is not the total of the column, as that would not be useful to anyone. It is the percentage of the workforce which is female. To calculate this, you divide the number of females by the total workforce, and mutiply by 100:

$$\frac{\text{Number of females}}{\text{Total workforce}} \times 100 = \frac{78}{188} \times 100 = 41\%$$

So, 41 per cent of the workforce is female.

See if you can complete Table 10.1 with the number of males in each department (this is the total of the department, less the number of females), and the percentage of males in each department. The first two have been done for you. Notice that the percentage of females plus the percentage of males, equals 100 per cent (in the Sales Department, for example, 55 per cent are female, 45 per cent are male - total 100 per cent).

Finally, calculate the percentage of males in the whole workforce.

See Feedback section for answer to this activity. Check your answers carefully.

Find three tables in use in your workplace. Do they show information in a clear and logical manner? Could they be improved? Is there any information which you currently use which could be put into a table to provide a clearer picture?

Grouping information

Some figures are easier to absorb if they are grouped in some way. Suppose you want to analyse the number of telephone calls in a month, made by different members of staff.

The numbers could range from, say, five for one person, up to several hundred for another. If you have 100 employees to count, producing a list showing the calls made by each person is not really informative. It would give a good enough picture of the calls if they were grouped together in a reasonable manner, for example:

Number of calls	Number of staff
20 and under	5
21-40	12
41-60	25
61-80	50
81-100	120
etc.	

Of course, you cannot tell from this grouping exactly how many calls were made. In the category 41-60 calls, all 25 people might have made only 41 calls each, or 60 calls each.

Averages

Sometimes it is useful to work on averages rather than on exact figures. For example, you might want to determine the average number of days staff were absent through sickness last year. To do this, you need to gather together data for each member of staff. Suppose you discover the following:

Name	**Days sick**
Peter	14
Jonathan	4
Mary	10
Judith	4
Tasleem	4
Bernard	8
Clifford	6

The arithmetic mean

The total number of days sick was 50. There are seven members of staff. So the average could be calculated as follows:

$$\frac{\text{Total days sick}}{\text{Number of staff}} = \frac{50}{7} = 7.14\,\text{days}$$

The average number of days off sick was 7.14 days.

There are two problems with this type of average - called the 'arithmetic mean'. The first is that it does not always produce a whole number as a result. Sometimes this is important. In this example, it is possible that a person could be sick for 7 days and a bit - but unlikely. People are usually sick for whole days, or perhaps half days. You have probably heard the joke about the average family having 2.4 children - how can anyone have part of a child?

The second problem is that the result does not give a figure equal to any of the absences listed above. No one was sick for 7 days. They were sick for either more or less than 7 days. So the figure of 7 days is not very useful.

The median

The median is the value in the middle of all the values. If you arrange the above sick days in order, you would get:

4 4 4 6 8 10 14

The value in the middle is 6. You could say that is the average number of days off sick, because three people have less sickness, and three people have more. At least it gives a number which exists in the list, but it still does not give a very accurate picture.

The mode

The mode is the number which occurs most often in the list. In this case it is 4 days. Three people had 4 days off sick, so you could interpret that as being the most likely period of sickness.

Imagine you are managing a shoe shop, and on a particular day, the following shoe sizes are requested by customers:

4, 2, 4, 6, 12, 2, 3, 6, 6, 6, 5, 6, 6, 12, 6, 2, 6, 5, 5.

Calculate the mean, median and mode shoe size requested. If you were only able to stock one size, which would it be?

See Feedback section for answer to this activity.

Index numbers

You have perhaps heard of the 'index of retail prices' which is used by the government to determine inflation and other information. There are many indexes produced by different organizations – the Financial Times Share Index, for example. The government also produces indexes which are published in the Annual Abstract of Statistics.

An index number is a comparison of one figure with another, with the first figure being converted to 100.

Say that you earned £200 a week in 1995, £220 a week in 1996, and £250 a week in 1997. Taking 1995 as the starting point, you first convert the actual value to 100, which represents 100 per cent of the 1995 earnings. This is called the 'base period'. The other 2 years are then also converted to percentages of the base period. To do this, you divide the actual amount in the year concerned by the actual amount in the base period and multiply by 100:

$$\frac{\text{Actual amount in year XX}}{\text{Actual amount in base period}} \times 100 = \text{index number}$$

So for 1996, the index number would be

$$\frac{220}{200} \times 100 = 110$$

In 1997, the index number would be

$$\frac{250}{200} \times 100 = 125$$

This tells you that the percentage increase in earnings between 1995 and 1997 was 25 per cent.

If you were to do this for your salary up until the year 2000, the index might appear as follows:

Year	**Wage (£)**	**Index**
1995	200	100
1996	220	110
1997	250	125
1998	300	150
1999	340	170
2000	360	180

So by the year 2000 your salary would have risen by 80 per cent compared with 1995.

An index can go down as well as up. If you took a cut in hours in 2001, such that your salary dropped to £240, the index for 2001 would be 120.

Produce an index for the following data regarding expected production in a factory:

Year	Tons produced
1995	1200
1996	1400
1997	1300
1998	1500
1999	1600
2000	1500

See Feedback section for answer to this activity.

Choosing the base year

The base year should be one which is stable and does not contain any unusual events. In the example above using salaries, there would be no point in choosing a year in which your earnings fell due to sickness.

Changing the base year

You do not need to stick to the original base year for ever. The purpose of index numbers is to clearly show the percentage increase since the base year. If the index numbers get too high for this to be clear, then the base year can be changed.

Suppose you have the following index of sales income:

Year	Sales income (£000)	Index (1990 = 100)
1990	120	100
1991	125	104
1992	150	125
1993	180	150
1994	250	208
1995	320	267
1996	400	333

It is clear that the original base year is no longer suitable, so it might be sensible to start again with 1996 as the new base year. The index would then continue as follows:

Year	**Sales income (£000)**	**Index (1996 = 100)**
1996	400	100
1997	440	110
1998	450	113
1998	470	118
1999	500	125

You can also re-state the earlier years if need be. The method is still the same, i.e.:

$$\frac{\text{Actual amount in year XX}}{\text{Actual amount in base period}} \times 100 = \text{index number}$$

So, 1995 would be re-stated as:

$$\frac{320}{400} \times 100 = 80$$

Activity 50

Calculate the re-stated indexes for 1990 to 1994, using 1996 as the base year.

See Feedback section for answer to this activity.

Examine a copy of the Annual Abstract of Statistics. You should find one in the library. It contains statistics about people, places, prices, types of industry and so on. Look at the various tables which affect your organization. Pick one or more of the indexes and see if the base year has been restated. If there have been significant fluctuations up or down, can you think of any reason why this might have happened?

Bar charts

Presenting information in the form of a bar chart is very common. The length of the bar shows the size of the items being considered. People often remember 'pictorial' information better than they remember lists of figures.

Consider the following information regarding the number of patients in a hospital:

Department	**Patients**
Casualty	20
Maternity	25
Geriatric	15
Medical	18
Surgical	30
Children	10
X-ray	24

Now close your eyes and try to say which is the busiest department, which is the least busy, and whereabouts did Maternity come in order of business? Unless you have cheated or are very good at absorbing figures you probably will not remember.

A bar chart, as in Figure 10.1, could show this information more clearly.

You can see at a glance that Surgical is the busiest, Children's department is the least busy, and the Maternity department is the second busiest.

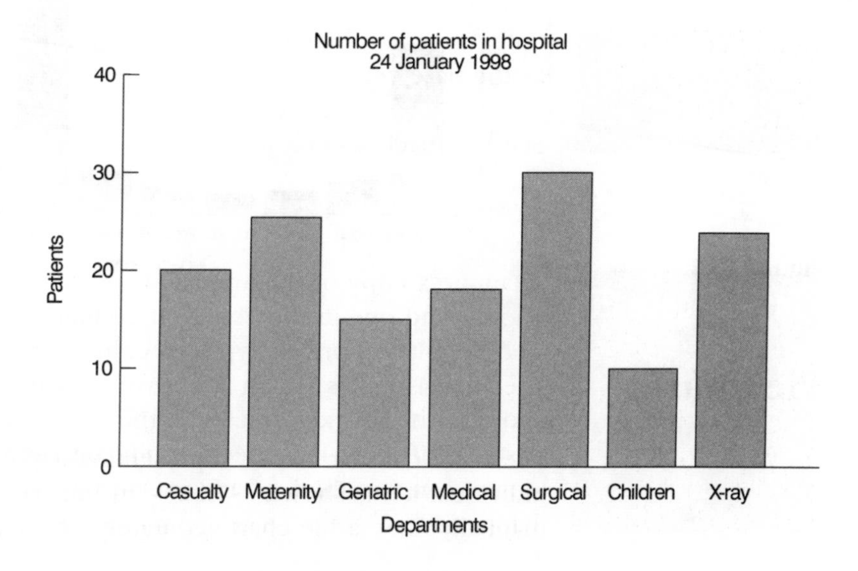

Figure 10.1

Most charts show quantities up the side and dates, departments etc. along the bottom, but there is no specific rule about this. If you drew this chart with quantities along the bottom and departments up the side, you would produce bars going across the chart, rather than upwards.

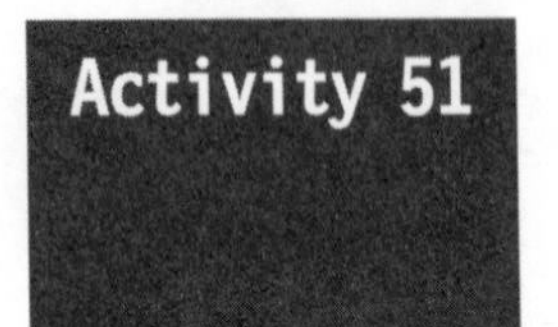

Activity 51

Draw a bar chart for the breakdown of total staff between departments from the information you were given on page 130.

See Feedback section for answer to this activity.

Bar charts can be presented in many different ways. If you are using a computer spreadsheet package they can be produced automatically for you, otherwise you have to draw them by hand.

Using the same example from page 130 again, Figure 10.2 gives a bar chart which shows the split between male and female employees, as well as the total.

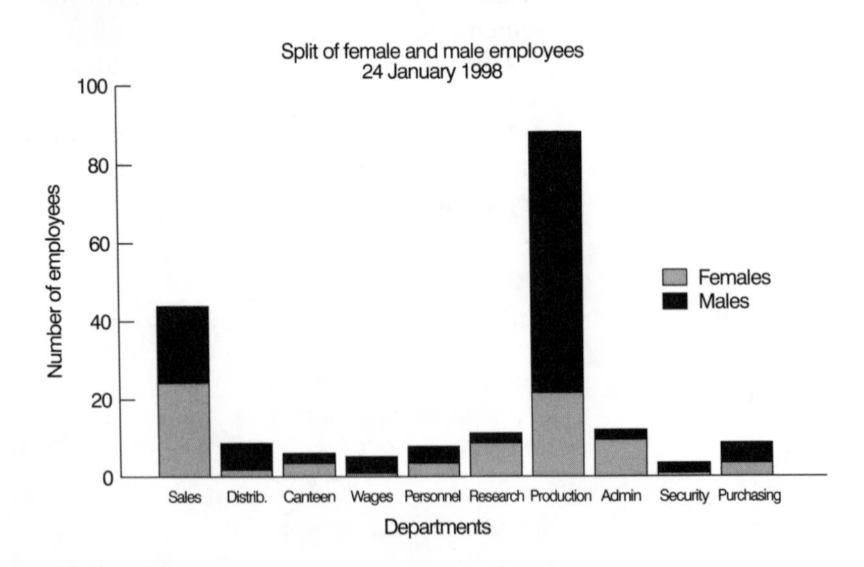

Figure 10.2

Pie charts

These are circular charts, representing a 'pie', with the individual items shown as 'slices' in the pie. In order to manually draw a pie chart accurately, you will need a compass and protractor to measure the size of the slices. However, it is unlikely that you will be called upon to hand-draw a pie chart, so it is assumed that you have access to a computer package which will do it for you.

Figure 10.3 shows the pie chart for the patients in the hospital, from page 139.

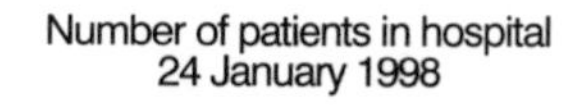

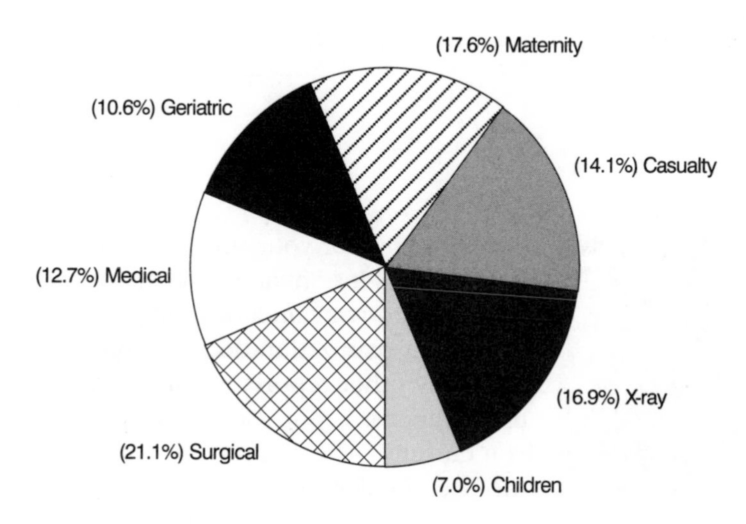

Figure 10.3

Line graphs

Graphs can be used for a variety of purposes, but a common use is to display how things have changed over a period of time. The graph is drawn by plotting a series of points on it by making dots or crosses and joining them up. Most line graphs have values up the side and dates etc., along the bottom.

So, from the data about salary, the first cross would be made by reading upwards from 1995, and across from £200 and placing a cross at that point. The next cross would be made at 1996 and £220 and so on.

In Figure 10.4 the graph shows the changes in salary from the data on page 136.

You can see that the salary rises annually.

Figure 10.4

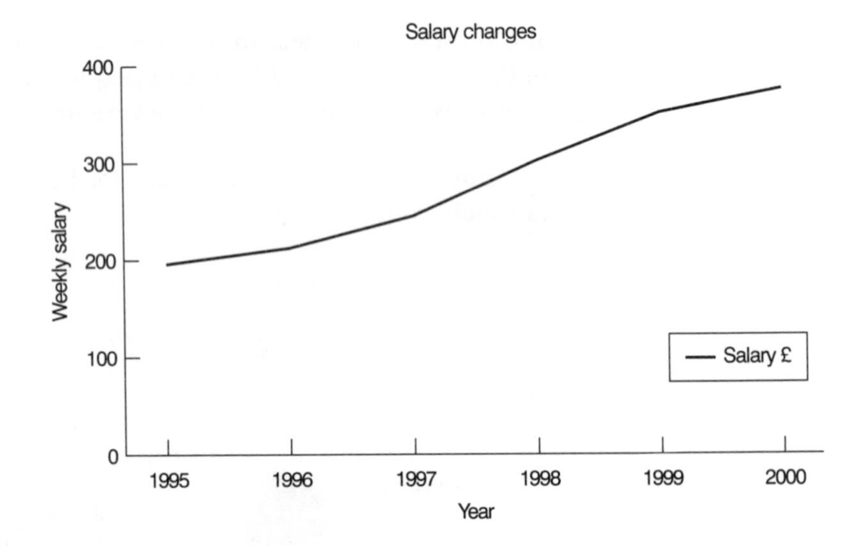

Obtain a copy of your organization's annual accounts or report. Does it contain any charts or diagrams? Are they a useful aid to your understanding of the figures in the report?

If your organization does not produce such a report, are there any other reports produced, perhaps internally, which use diagrams? If not, can you think of any information which you currently use or produce which could be made clearer by the use of diagrams?

Identifying trends in information

Information is provided to aid decision-making. If you sense that something is happening in your organization which might need action, then you should investigate and take that action or report it to the appropriate person.

But in today's organizations, it is not sufficient to rely solely on your 'sense' of things. Evidence is also required to back up your opinions.

Some events are 'one-off' events, for example, if an employee has a car accident and is away from work for 6 weeks. This does not require the same action as an employee who is regularly off once a week. This would be noticed by examining working patterns over a period of time. Any changes which occur, or connections which can be made, can point to patterns of activity which might require action.

There are two main types of pattern which you, as a team

leader, are going to consider. One is how things change over a period of time (known as 'time series analysis') and the other is how things change in response to other things (known as 'correlation').

Time series analysis

This looks at how things change over a period of time. It is very useful for determining what might happen in the future. The statistical techniques involved in time series analysis are quite complex and are beyond the level of knowledge you would be expected to have as a team leader.

It is, however, useful to look at the diagrammatic method of identifying changes over time.

Imagine you are a team leader in a firm of solicitors. There are four solicitors involved in property conveyancing (buying and selling of houses). Sometimes they seem busier than others with such work. When they are very busy, the demands on the secretarial staff whom you manage, are high and you sometimes have to bring in temporary staff. At other times, there seems to be less work to do. It might be useful to plot the workload on a graph.

Suppose the figures over the last 4 years are as follows:

Year	Quarter	Number of conveyancing jobs completed
1994	1	120
	2	230
	3	250
	4	140
1995	1	130
	2	250
	3	280
	4	150
1996	1	140
	2	260
	3	300
	4	180
1997	1	120
	2	270
	3	320
	4	200

If each of these quarters' jobs are plotted on a graph, they appear as in Figure 10.5.

You can see that the number of jobs dips in the 1st and 4th quarters every year and is higher in the 2nd and 3rd quarters. Apart from the 1st quarter of 1997, the number of jobs is increasing - so you can say that there is an 'upward trend'. This could help you to determine the need for additional staff in the future.

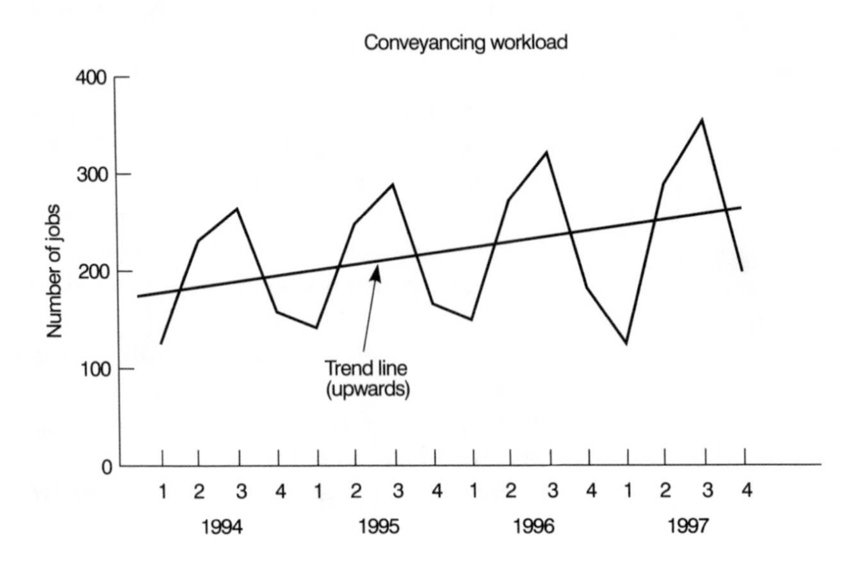

Figure 10.5

The 'trend' can be seen by putting a ruler on the chart, around the middle of all the various lines, and determining whether it slopes up or down.

You can see that the trend in this case is upwards - the number of jobs is increasing overall.

The ups and downs are called 'seasonal variations'. In this example, we have used the four quarters of the year, but you can prepare this kind of chart for any time series.

Draw a time-series graph for the following data regarding the numbers of staff off sick during a month:

Week no	Day	No. of staff absent	Week no	Day	No. of staff absent
1	Monday	10	3	Monday	15
	Tuesday	2		Tuesday	4
	Wednesday	3		Wednesday	6
	Thursday	4		Thursday	8
	Friday	6		Friday	10
2	Monday	12	4	Monday	18
	Tuesday	3		Tuesday	4
	Wednesday	5		Wednesday	6
	Thursday	7		Thursday	7
	Friday	9		Friday	12

See Feedback section for answer to this activity.

Correlation

Correlation is where one thing changes in response to another.

Consider an organization trying to decide whether advertising has any effect on the sales of the business. Suppose it has the following data for the last year:

Month	Advertising costs (£000)	Sales revenue (£000)
January	10	2000
February	12	2500
March	10	2200
April	15	3200
May	12	2600
June	15	3000
July	18	4000
August	20	4500
September	15	2800
October	12	2000
November	12	2300
December	10	2100

Looking at this table of data, you can see that there are three months in which the firm spent £15 000 on advertising. The sales revenue which resulted was £2800, £3000 and £3200 in the three months. This tells you that there is not a perfect relationship between the amount spent on advertising and the sales revenue - otherwise the same expenditure would result in exactly the same revenue. However, you can probably see that lower expenditure on advertising tends to produce lower sales revenue and higher advertising produces higher revenue.

You can plot this data on a graph. This type of graph does not need to mention the time period - the two things you are comparing are 'advertising expenditure' and 'sales revenue'.

Place your dots or crosses for each pair of figures. This is called a 'scatter diagram'.

The diagram would look like the one in Figure 10.6.

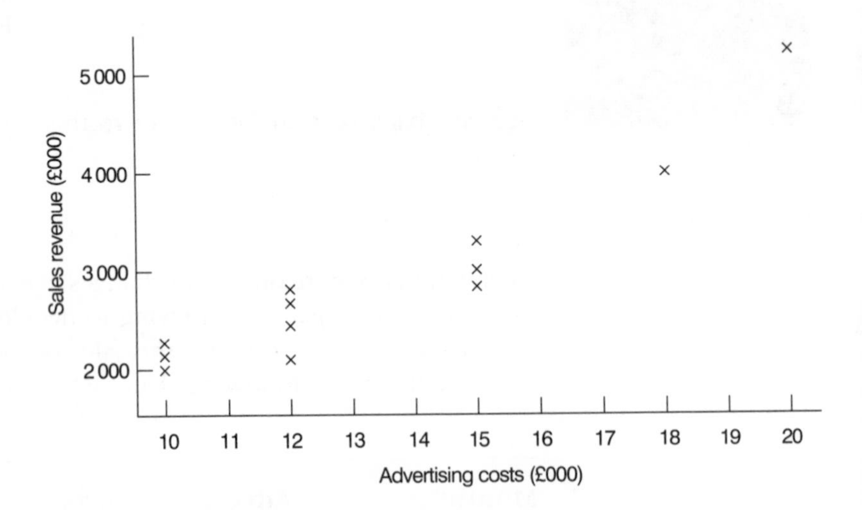

Figure 10.6

The diagram in Figure 10.6 shows that there is a relationship between the amount of advertising and the sales revenue which results, because as one increases so does the other. This is called 'positive correlation'. Negative correlation occurs where one moves up as the other moves down. An example of where this might happen is if you increase the amount spent on staff training and the rate of absenteeism decreases.

It is important not to draw too many conclusions from these relationships. It could be, for example, that as the number of invoices received by an organization rises, so does the number of staff who use the after-work sports activities. It is very unlikely that the two are related in any way.

Summary

This chapter has considered the presentation and interpretation of information. It is important that information is presented in the way in which it is best understood by its users. Often pictorial methods are easier to absorb than lists and tables of figures. It is also important that the information presented is interpreted correctly. Remember that not all data is totally reliable, depending on how it has been collected and input.

Nevertheless, the methods examined in this chapter should enable you to look at alternative ways of putting across your message in the future.

Review and discussion questions

1 Do you think that tables are a good way of presenting information?
2 What are the three ways of calculating an 'average' from a list of figures?
3 What is an index number and why is it useful?
4 Describe four types of chart.
5 What is a time series analysis? How can it be used to identify a trend?
6 What is meant by correlation?

Case study

Dennis is a foreman with a firm of builders. He is responsible for the costing and pricing of jobs below £20 000, and the hiring of sub-contractors for particular jobs. He has experienced the following problems recently:

1 The cost of building materials has been rising recently. He has been using last year's costs and adding a small percentage to them to allow for inflation. However, some materials have risen more than others and this has sometimes resulted in incorrect calculations of the cost of jobs.
2 The majority of jobs involve costs for basic materials, labour, plant hire and other items. He sometimes finds it difficult to pull all these costs together to determine an overall cost.
3 The proprietor of the firm is a good salesman, but does not grasp the meaning of figures very quickly. He is constantly asking Dennis for a breakdown of costs for individual jobs and when Dennis gives him the figures he seems not to absorb them fully.

4 Dennis has noticed that jobs do not come in on a regular basis. There seems to be a greater demand in the spring and less in winter, particularly for outside jobs. He finds it difficult to determine how busy the firm is going to be from month to month.

Identify and describe various methods of presenting and interpreting information which Dennis might find useful to fulfil the above requirements.

Identify what tables, charts and diagrams are used in your workplace. Are they clear and helpful? Could any of them be improved? Can you identify other areas where the information given could be presented in a more understandable format using a chart or diagram?

For your own particular job, design a table to analyse the different tasks you undertake during a month. Restrict the table to no more than ten tasks. For each task, fill in the number of times you perform it, and the length of time it takes. If you perform it several times, calculate an average time spent on the task (this might mean keeping a separate record of the different occasions on which you carry out the task).

Produce a suitable diagram or chart which shows the breakdown of tasks carried out during your working month.

11 Undertaking a project and writing the project report

Learning objectives

On completion of this chapter you will be able to:

- select an appropriate subject for your project
- write your project terms of reference
- plan your project
- investigate the subject of your project
- analyse the information you have gathered
- reach conclusions and make appropriate recommendations
- write a project report which meets the NEBSM Certificate in Supervision and Management guidelines.

Introduction

When we looked at report writing in Chapter 8, the emphasis was on writing a work-related report. If you are studying for a supervisory management qualification you will probably have to do at least one project and write a project report. If you are studying for the National Examining Board in Supervision and Management (NEBSM) Certificate you will definitely have to undertake a project and write a project report. This chapter has been designed to help you do this. The example of the NEBSM report has been used throughout, but the principles are the same for any work-based project report that you have to present for assessment, if you are working towards a supervisory management qualification.

You will be able to use many of the techniques in this chapter during your study of *Resources Management*.

Why undertake a project?

Undertaking a project enables you to bring together the many things that you have learned during your course of study and demonstrates how you can apply your learning to a real situation. The project is the activity that you undertake, the report is the written work that you produce at the end of your project. If you are studying at college your tutor will support and advise you whilst you are working on your project, this tutor role is often called project supervision. Your supervisor will brief you on the requirements of the project, help you to define the terms of reference (TOR) and, as you progress, comment on drafts of the project. Your supervisor may also visit you at work to discuss your project with you and advise on progress.

The project tests your ability to observe and record facts relating to the problem/issue that you are investigating. It also tests your ability to analyse the information that you have collected which will lead to your conclusions and recommendations. The writing of your project report demonstrates that you can present a convincing argument in writing.

Choosing a project

NEBSM require you to undertake a practical work-related project. It must be based on problem-solving in the workplace. You need to select a problem at work, investigate the problem, and make some recommendations about how you believe the problem can be solved. Figure 11.1 is an extract from the NEBSM project guidelines giving guidance on some of the essential elements of a NEBSM project.

The NEBSM guidelines state that it is essential that your project should be:

- concerned with a problem of a supervisory/management nature and not one which requires a purely operational/technical solution
- defined by clear terms of reference, setting out the boudaries of the investigation before the start of any detailed investigatory work
- based on problem-solving in the workplace and require the candidate to undertake investigative and analytical work, to draw conclusions and to make relevant recommendations.

Reproduced by permission from NEBSM Project Guidelines

Figure 11.1 NEBSM project guidelines

In addition to the above, it is vital that your project demonstrates an *awareness of financial issues*. You need to be sure that any recommendations that you make are likely to be financially viable and acceptable to the organization. Your project should also include reference to the *personnel/human factors*. For example, if you are proposing changes that affect peoples' jobs, it is important to consider all the human resource implications, such as how you propose that staff should be consulted or what type of training would be required, how much it would cost and how it could be organized.

To summarize then, your project must:

- deal with a supervisory/management problem
- be defined by clear terms of reference
- be based on solving a problem in the workplace
- demonstrate an awareness of financial issues
- include reference to human resource factors
- include investigative and analytical work, which leads to conclusions and recommendations

A work-based project of this type will be a useful learning experience for you and will also produce benefits for your employing organization. Your project will provide suggestions as to how your employer might solve a problem that might, for example, increase efficiency, improve quality or save money.

It is important to discuss your choice of project with your line manager, who may have some suggestions about a topic for your project and will be able to provide you with help and support whilst you are working on your project.

You need to have very clear terms of reference. Your line manager should agree your terms of reference, as should your project supervisor and NEBSM External Verifier. Your project supervisor will arrange to send your terms of reference to the External Verifier for approval.

The terms of reference should be very specific. They must set out exactly what you are investigating and what the outcome of your investigation will be. For example, to make recommendations, to devise a system for, or to make suggestions for improvement, Figure 11.2 is an extract from NEBSM guidelines on terms of reference.

It is a good idea to include in your terms of reference the fact that you will be dealing with the financial and human resource implications of the problem. This will ensure that you cover these essential elements in your report. Figure 11.3 shows some examples of terms of reference.

The NEBSM guidelines say that the terms of reference should include:

- a clear definition of what is to be investigated, including the boundaries
- the purpose of the investigation
- the need to draw conclusions and make recommendations (perhaps taking account of specific constraints).

Figure 11.2
NEBSM guidelines on terms of reference

Reproduced by permission from NEBSM Data Sheets

- To examine the present system of analysing and meeting the training needs of staff working in the packing section.
- To determine if the current system is appropriate to meet the needs of the staff and the organization.
- If the current system requires improvement, to make costed recommendations for the introduction of a new system which would be acceptable to the staff and the organization.

To establish why, at the Swansea depot, there are substantial discrepancies occurring between the actual stock of finished goods, compared with the stock holdings showing on the computerized stock records.

To examine the current paperwork and procedures relating to both physical and computerized stock holdings.

To make fully costed recommendations that will, if implemented, improve the stock recording and control system and reduce the discrepancies. The project will include an analysis of the impact that the recommended changes would have on the people working in the organization.

1 To investigate increases in the numbers of customer complaints over the period from January 1995 to December 1997.
2 To establish why the number of customer complaints has been increasing.
3 To make fully costed recommendations, which take account of human resource issues, that if implemented will result in a reduction in customer complaints.

Figure 11.3
Examples of terms of reference

You are now ready to decide your terms of reference.

Check your terms of reference against the checklist below. Make sure that your terms of reference have these features:

- they are clear and concise
- they indicate that the project will deal with a supervisory/management problem
- they are based on solving a problem in the workplace
- they indicate that you will demonstrate an awareness of financial issues
- they indicate that you will include reference to human resource factors
- they indicate that your project will include investigative and analytical work leading to conclusions and recommendations.

Planning your project

Once you have decided your terms of reference you need to sit down quietly and do some planning. You need to plan your project very carefully. The best way to do this is to produce a project calendar, which is your working document for you to plan ahead and check your progress against your plans. Figure 11.4 shows a sample project plan.

When constructing your plan, do not forget to take into consideration practicalities such as, it is unlikely you will do any work on your project over Christmas or when you are on holiday.

You also need to think about how much time you will need to spend either at work when you are conducting your investigations, seeking out information and talking to people and at home where you will probably do much of your thinking and writing. Ensure that your manager and family know that you will need the time to do this work.

Presentation of the report

The NEBSM project report needs to be about 3000 words. The report should be word processed (or typed) on A4. If you do not have access to facilities to enable you to do this yourself, remember to build time in your plan for someone else to type your project. You will also need to check it through and return it to have the inevitable mistakes put right.

Week no	Key milestones	Details of actions to be carried out each week	Working notes on progress
1		Think about TOR. Discuss with line manager	
2	Hand in TOR to project supervisor	Agree TOR with line manager and project supervisor	
3		Do research	
4			
5			
6	Complete investigations		
7	**HOLIDAY**	**HOLIDAY**	**HOLIDAY**
8		Write Section 1 – Introduction	
9			
10		Write Section 2 – The Present Situation	
11			
12		Write Section 3 – Investigation	
13			
14		Write Section 4 – Analysis	
15			
16		Write Section 5 – Conclusions	
17			
18		Write Section 6 – Recommendations	
19	**HOLIDAY**	**HOLIDAY**	**HOLIDAY**
20		Write Summary	
21	Completed draft to project supervisor		
22	Draft handed back to me	Make any alterations and do any further work suggested by project supervisor	
23			
24		Sort out Appendices	
25		Write Bibliography	
26		Organize Front Cover, Title, Acknowledgements, Contents and Terms of Reference pages	
27		Put it all together and finish off	
28		Project to line manager for comment	
29			
30	Project to be handed in to NEBSM tutor		

Figure 11.4 A project plan

The pages of your report must be numbered and it should be split into sections, each section having a section heading. See Figure 11.5 for an appropriate report format.

- Front cover
- Title page
- Acknowledgements
- Contents
- Terms of reference
- Summary
- Introduction — Section 1
- The present situation — Section 2
- Investigation — Section 3
- Analysis — Section 4
- Conclusions — Section 5
- Recommendations — Section 6
- Appendices
- Bibliography

(Sections 1–6: 3000 words, 10–12 pages)

Figure 11.5
NEBSM report format

You are now ready to complete your project action plan. Use the following pro forma to help you organize your time.

Week no	Key milestones	Details of actions to be carried out each week	Working notes on progress

When you have completed your planning you are ready to get yourself organized in a practical sense. You will need one place to keep all your project work together. You will need a notebook to jot down notes as you work, an A4 pad for writing the drafts of your project and a ring binder to keep your work in progress together in one place.

Your final report should be presented in a binder which protects it and enables the reader to turn the pages and read the report easily. It is usual to prepare multiple copies of a project report. For the NEBSM project report you should produce three copies, one for yourself, one for your line manager and one for your NEBSM centre, which is the copy you hand in to your project supervisor.

You should include charts and graphs where these usefully illustrate a point you are making and discussing in your project. For example if you are showing production figures or analysing questionnaire results, this type of information is much easier for the reader to understand provided in the form of charts and graphs. You can also use tables to compare information. For example if you are comparing the costs of various solutions to a problem it might be a good idea to present the information in a chart so that it is easy for the reader to compare. You can also include photographs in your project if they help to illustrate a point. In most cases this type of information should be included in the appendices. The flow of your writing is then not broken up. For example,

'50 per cent of the sample that completed the questionnaire had not received any training on the task; see appendix 3 for an analysis of the questionnaire results.'

Appendix 3 would be a chart illustrating the responses to the questionnaire.

Make a list of the graphs, charts and illustrations that you will include in your report. (You have covered how to produce this kind of information in Chapter 10.) Your report will be much more interesting if you use visual material. It will also demonstrate to the person marking your project that you can use a variety of techniques to communicate information. Ask your project supervisor to check the list that you have compiled.

Project reports should be typed in double spacing on one side of A4 paper. The left hand margin should be 38 mm (1 1/2 ins) and all other margins (top, bottom and right side) 25 mm (1 in). Using size 12 type, with the appropriate margins and spacing, each page will have 25 to 30 lines (about 250 to 300 words). Your finished project, therefore, will be 10 to 12 pages long, from the introduction to the end of the recommendations (see Figure 11.5).

You should now have decided upon your terms of reference, completed your project plan, know what graphs, charts and illustrations you will include, and have collected together the materials that you will need whilst working on your project. In addition you can now visualize the finished project and so will have a good idea of the size of the task ahead. You are now ready to begin your investigations.

Conducting the investigation

Your investigation will provide the material that you will use to write sections 1 to 3 of your report: Introduction, The Present Situation and Investigation. In the subsequent sections of the report, you will go on to analyse the information, reach conclusions and make recommendations.

Your terms of reference will indicate that there is an issue or a problem that you are investigating. In the early stages of working on your project you will be collecting information relating to the issue or the problem.

You will need to collect enough information for you to write the Introduction, where you will give some background information on your organization and describe the problem you are investigating and why you are investigating this particular issue. Give some background and details of any evidence you have that the investigation is necessary. Explain to the reader what you are aiming to change or improve by undertaking this project.

In section 2, The Present Situation, you will need to describe the current situation in relation to the subject of your project. This means that you will need to find out exactly what the situation is now.

Section 3 begins with a description of exactly how you carried out the investigation. In other words, how you have carried out your research. You also include the results of your investigation in this section. Remember, if you are using graphs, charts or questionnaires, to put the details in an appendix and to discuss or refer to the information in the main body of report.

You will have to consider carefully how you are going to carry out your investigation. Consider who you will need to talk to, how you are going to get the information you need and whether you need any help from others. Make sure that you keep all your project information together. This is crucial at this stage, as you will be collecting a great deal of information and it is important you do not lose anything and waste precious time searching for something you have mislaid.

You need to take care that you are basing all your investigation on facts and that the reader can find the factual information that you are referring to. This is best illustrated by using an example.

'The number of working days lost due to sickness has consistently increased over the last three years.'

The above statement must be supported with evidence. You need to provide the detailed evidence to support what you are saying. You would need to include information about the number of working days lost due to sickness over the last three years. Ideally the detailed data would be contained in an appendix, so readers can refer to it if they wish.

The improved sentence might read:

'The number of working days lost due to sickness has increased from 50 in 1995 to 100 in 1997. This represents an increase of 100 per cent. See appendix C.'

Now you have gathered the facts relating to the issue that you are investigating and the rest of the project will be based on these facts. It is important that you base your subsequent analysis, conclusions and recommendations on the facts you have discovered and nothing else. Do not introduce any new material at this stage.

Analysing information, reaching conclusions and making recommendations

In this section you will analyse the data that has resulted from your investigation. Obviously it is difficult to generalize about how you will analyse your data, as every project is different. Look for the general message: what is the information indicating? Does it indicate that matters could be improved if changes were to be made? Are there any obvious trends? In your analysis you might use some ideas and concepts that you have studied during your recent course of study. For example you may be able to relate a lack of motivation to some of the motivational theory you have looked at. This type of analysis adds depth to your project and shows that you can relate what you have learnt to the real world.

The analysis will lead to the conclusions. You will draw conclusions from the data you have analysed about what the situation is now and what needs improving.

Your recommendations will state what you propose should be done now. They will be based on your analysis and conclusions and on the evidence in your report. You must demonstrate that you understand the effects of your recommendations in both financial and human resource terms. You can include a separate statement giving information of costs and/or savings that would result if your recommendations were adopted.

Writing the report

You are now ready to write the report. You should write in simple English, in a straightforward business style. Your report will be in sections. Look again at Figure 11.5 to remind you of the sections of the report. Each section should be numbered and have a heading. If you are dealing with a number of issues within the section use numbered subheadings to break up the text and to help the readers of your report find information more easily. See Figure 11.6 for an illustration of how to use numbered headings and sub headings.

We will now examine each part of the final report in more detail.

1 INTRODUCTION

1.1 ABC LTD
ABC Ltd has been trading for 25 years. ABC Ltd is a mail order company dealing in . . .

1.2 The Packing Department
I have worked for the organization for 10 years and for the last 2 years I have been a Team leader in the packing department . . .

1.3 Background to the Problem
The business has grown very rapidly over the last 3 years, the directors have been concerned for sometime about . . .

6

2 THE PRESENT SITUATION

2.1 Method of working
The current method of working is . . .

2.2 Training
When staff join the organization they undertake a programme of induction. See appendix A for the detailed content of the current induction programme. The main areas covered in the induction training are . . .

8

Figure 11.6
Headings and sub headings

Front cover

This has a practical purpose, it will protect your project and, if it is neat and smart, create a favourable first impression. The front cover should give the title of the project and your name. The title of your project will be briefer than the terms of reference, just a few words that indicate what the project is about. See Figure 11.7.

Figure 11.7
The front cover

The title page

The title page should show, not only the title of your project, but also your name and job title as the author, the title of the qualification, the name of the awarding body, the month and the year. See Figure 11.8 for a sample title page.

Acknowledgements

It is usual for the author to thank those people who have helped and supported them whilst working on the project. Obviously you will wish to thank the people who have particularly helped you, but an example of an acknowledgements page is shown in Figure 11.9.

Figure 11.8
The title page

PROJECT TITLE	Project title in capital letters underlined
by	
YOUR NAME	Name in capital letters underlined
Team Leader	Your job title
Certificate in Supervisory Management	Title of qualification
National Examining Board in Supervision and Management	Name of awarding body
Stockport College of Further and Higher Education	Name of the organization to which the project is to be submitted
May 19xx	The month and the year of submission

ACKNOWLEDGEMENTS	
The author would like to thank the following people who provided practical help and moral support during the project.	A paragraph saying why you wish to thank people
Mr David Brown, Production Manager, ABC Ltd	It is usual to give the person's name and their job title
Ms Sheila Robinson, Personnel Manager, ABC Ltd	
Mrs Gill Humphrey, NEBSM Course Tutor, Stockport College	

Figure 11.9
A sample acknowledgement page

Contents

This page gives details of the contents of the project showing section numbers, section titles and page numbers. The contents page helps the readers to find their way around the report. See Figure 11.10.

CONTENTS

	page
Terms of reference	1
Summary	2
1 Introduction	3
2 The Present Situation	4
2.1 xxxxxxxxxxxxxx	4
2.2 xxxxxxxxxxxxxx	5
3 Investigation	6
4 Analysis	8
5 Conclusions	12
6 Recommendations	14
Appendices	
Appendix A xxxxxxxxxxxx	16
Appendix B xxxxxxxxxx	18
Bibliography	19

Figure 11.10
A sample contents page

Make a rough draft of your contents page. You can add the page numbers as you work on your project, but it will help you to have your contents mapped out now. Include the section numbers, headings and subheadings that you are proposing to use. You may change your plans as you go along. This is fine, but it is still worth planning now even if you change your mind later.

Terms of reference

We have already covered the writing of the terms of reference. It is a good idea to put these on a separate page so that they are easy for the reader to pick out and refer to.

Summary

This is the last section that you write. Write the summary after you have completed the rest of your report. This briefly says what the report is about and what the major conclusions and recommendations are. Write the summary last but place it at the beginning of the report. The idea is that the readers can quickly and easily read an overview of the content and grasp the main messages. The summary should be no more than one page.

Introduction

This section introduces the topic of the project. Include an introductory section giving a little background information on your organization, your section and your job role. This helps to put the project in context for the reader. It is a good idea to describe the problem that you are investigating in this section. Explain why this is the topic that you have chosen for your project. Explain the background to the problem, what the situation is now and why the current situation creates a problem. Say what you are aiming to change or improve by undertaking this study. It might also be sensible to define the parameters of the project here, that is any assumptions or limitations that you had to work within, any aspects that you have not covered.

The present situation

Describe the current situation, in relation to the subject of your project. To do this you will have to investigate exactly what the situation is now.

Investigation

Firstly, describe how you have carried out your investigation into the problem. Explain the method that you have used. For example who have you talked to? Have you conducted any questionnaires? Describe exactly how you undertook your research. Also describe any problems that you encountered when you were carrying out your investigations, for example, if you did issue questionnaires, was the return rate poor? Did you have problems getting hold of some of the information that you needed?

Include details of the information that you gained. If you are using graphs, charts or questionnaires place the details in the appendices but refer to the information that you found out in this section. (See also 'Conducting the investigation' on page 157.)

Analysis

See 'Analysing information, reaching conclusions and making recommendations' on page 158. In this section you will analyse the data that you have gleaned during your investigation.

Conclusions

You will draw your conclusions from your analysis. The conclusions do not contain any new factual material. The conclusions could include reference to what the overall situation is now and what will/might happen if nothing is done about it. Then you might consider a number of solutions, possible courses of action and the costs, problems and benefits of each course of action, before taking a decision on exactly what you are going to recommend.

Recommendations

These are your suggestions based on your investigations and conclusions about what should be done next. What course of action do you recommend to solve the problem that you have investigated? Recommendations should be brief. Deal with each recommendation in a separate numbered paragraph. Ensure that your recommendations are based on the evidence you have given in the main body of the report. Ensure that you have made clear exactly what you think should be done and that you understand the effects of your recommendations; remember that you have to demonstrate that you understand the implications of your recommendations in relation to financial and human resources. You can include a separate statement in this section detailing the financial implications of your recommendations. This would show the costs and/or savings that would result if your recommendations were implemented.

Appendices

This section should contain the supporting evidence that you discovered in your investigation. You can label your appendices A, B, C and so on or 1,2,3 etc. or with Roman numerals I, II, III, IV etc. It is not appropriate to include detailed information in the main body of the report so put any supplementary information in the appendices, such as graphs, charts, detailed calculations, diagrams. If you use appendices properly, you will not clutter up the main report with too much detail. It is important to refer to the appropriate appendix in the main report. This will enable the reader to find supporting information quickly and easily. Do not use the appendices as a dumping ground for information that you are not sure where to put. Make sure that every appendix is relevant and necessary. If you are using a number of technical terms in your report include a glossary of terms as the first appendix.

Bibliography

Give details here of any books or articles which you have used as a reference whilst writing your project. You should provide enough information for the readers to trace the original material if they want to.

Final check and follow up

You are nearly there now! Check the report for mistakes and to make sure that it still meets your original purpose. It is essential that you follow up the report to try to ensure that appropriate action is taken. Use the checklist in Activity 57 to ensure that you have covered everything that you need to.

Activity 57

Checklist for your completed project.

When you are at the final draft stage, use the checklist below to make sure that you have covered everything.

Area	Item to be checked	✔
Presentation	Is it well presented and laid out?	
	Is there a useful table of contents, a summary, a main theme with conclusions, recommendations and appendices?	
	Is it likely to stimulate action?	
Investigation	How was it researched and how much detail is included?	
	Are the facts supported by the evidence?	
	How were problems overcome?	
Relevance	Is all material relevant to the terms of reference?	
Completeness	Has adequate information been collected?	
	Have all aspects, including finance and human factors been covered?	
Charts	What charting and graphic skills have been used?	
	Do they support the written word?	
Conclusions	Are they realistic?	
	Are they backed up by the information in the body of the report?	

Checklist adapted from NEBSM Data Sheets.

Summary

In this chapter we have looked at how to produce a project that will fulfil all the requirements of the NEBSM Certificate in Supervisory Management project. If you follow all the guidelines in this chapter you will produce a quality project that covers all the criteria NEBSM require.

It will be hard work and there will be times when you wonder why on earth you ever started your programme of study. On the other hand, when you complete your project, you will have a great sense of achievement and will have produced a report which demonstrates how much you have developed your skills whilst undertaking a course of study and that has produced benefits for your employer. It cannot be over emphasized enough how important it is to plan well in advance and keep yourself as organized as possible throughout. If you are not studying for a NEBSM qualification, much of the content of this chapter is still very relevant for you as it can apply to any work-based project that you need to complete as part of a programme of study leading to a supervisory management award.

Feedback

Activity 2

You might have said that it is expensive to provide, time-consuming to provide, and must lead to correct decisions being made.

Activity 3

You could include items such as:

- the address of a customer who needs goods by 3 pm this afternoon
- the blood group of an injured person needing an immediate transfusion
- the time of trains to London this evening
- the last date for the return of goods bought on 'sale or return'
- the production of an agenda for a meeting.

Activity 4

Formal information will include the receipt of forms, documents, letters, memos, etc., while informal information will be oral or visual.

Activity 5

Statements 1, 3 and 4 are qualitative; Statements 2, 5 and 6 are quantitative.

Activity 6

Characteristic of information	Leisure pool	Discounted admissions	Lifeguard rotas
Source	External - surveys of customers, competitors, costs	External and internal - prices at other centres, own prices, lost profit, estimate of increased use	Internal - expected use of pool, availability of staff, maximum hours to be worked
Scope	Very wide, not defined	Short and medium term effects	Narrow, well-defined - internal only
Level of detail	Summarized costs and revenues	Summarized monthly figures and comparisons	Hourly rotas, different for each day
Currency	Future	Past experience, present and future demand	Past experience, present needs
Timeliness	Several months hence	Some urgency, before new school year	Immediate - for next week
Frequency	Infrequent - never been done before!	Every year or term	Frequent - done regularly, amended daily if absences
Level of precision	Low, approximate, qualitative	Fairly precise costings; estimates of results	Exact times

Activity 7

Staffing over Easter would involve the following:

- *Internal information* - regarding staff available, overtime rates, holidays booked, staff sick
- *Well-defined information* - staff willing to work, hourly rates of pay, days available, opening hours
- *Detail* - individual names, rates of pay, working hours
- *Past information* - existing staff, current pay rates, previous willingness to work
- *Immediate information* - needed in time to notify staff and make decisions before Easter
- *Frequency of information* - same decision has been made several times before
- *Exact information* - to the nearest hour or half-hour, to the nearest pound.

Activity 8

Items 1,3 and 6 are examples of data. Items 2, 4 and 5 are examples of information.

Activity 9

ABC Mail Order Company
High Street
Manchester, M1 3EJ

0161 234 5678

ORDER FORM

Order number: ____________

Date: ____________

Agent: ____________

PLEASE COMPLETE IN BLOCK CAPITALS AND PRESS FIRMLY.

Catalogue no.	Page	Description	Quantity	Price each	Total price
		TOTAL VALUE			£

Customer name ____________ Customer number ____________

Customer address ____________

____

____

Telephone number ____________

DO NOT ENCLOSE ANY PAYMENT WITH YOUR ORDER. YOU WILL BE NOTIFIED OF THE METHOD OF PAYMENT WHEN YOUR GOODS ARE DELIVERED BY YOUR AGENT.

PLEASE CONTACT YOUR AGENT IF YOU HAVE ANY QUERIES.

THANK YOU FOR YOUR CUSTOM

Activity 10

You probably chose the following methods:

1 questionnaire
2 observation
3 completing forms
4 interview.

Activity 11

A *data subject* is a living individual who can be identified by the data which is stored about them on a computer system. A *data user* is someone who stores, processes or controls data about data subjects, on a computer system.

Activity 12

Well, all except item 3 would require registration. Item 3 is required only for accounting and general business purposes.

Activity 13

You might include the slowness of maintaining records, the difficulty in extracting information from them, the space they take up, the need for duplicated records for different purposes and the chances of error.

Activity 14

The four files are master, transaction, reference and program files. In an order processing system the master file would be the file of wages paid to date; the transaction file would be the hours worked this week or month; the reference file would be the file giving tax codes, tax rates etc.; the program file would be the file which instructs the computer how to calculate the wages.

Activity 15

Examples of input devices include keyboard, screen, barcode reader, character reader, mark sensor, document scanner, voice recognition, magnetic strips and computer-to-computer.

The advantages of keyboard are that it is familiar to use, easy to obtain and cheap. The disadvantages are that it is slow and prone to error.

The advantages of bar code readers are that they are fast and accurate and can be used to access the computer to obtain details of prices, product descriptions, etc. and to automatically update the accounting and stock records. The disadvantages are the cost of putting bar codes on items and the cost of the equipment used.

The advantages of character readers are that they can automatically read text and numeric data by recognizing its shape. There is no need to enter data manually so the error rate is low and entry is fast. They can also be used as 'turnround documents' when produced by the computer and used to feed back into the computer with new data. The only disadvantage is the cost of printing the characters and the equipment needed to recognize them.

Activity 16

Printers, screen, microfilm and microfiche, graph plotter, voice output, discs and tapes.

The advantages of printers are that they are cheap, produce high quality hard copy, perhaps in colour. The disadvantages are that they can be noisy and the cost of paper storage is high.

Activity 17

The most suitable piece of software for each of the activities is:

- cash budget - spreadsheet
- report - word processor
- customer details - database
- advertising leaflet - desk-top publishing package
- presentation - graphics package.

Activity 18

You could choose from accounting, payroll, stock control, purchasing, invoicing, costing, personnel and many others.

Activity 20

Your answer should include Transaction Processing systems, Reporting systems, Decision Support systems, Executive Information systems, Expert systems and Design systems. For details of each of these, re-read the paragraphs above this activity.

Activity 22

Factors to ensure a smooth changeover to a computerized system would include the following:

- involve users at all stages
- ask the advice of users
- encourage user involvement in the development of the system

- emphasize the benefits to the users, e.g. better information, easier to operate, improving their own skills
- introduce changes gradually
- support staff with help and reassurance
- give adequate training
- maintain existing working relationships
- choose user-friendly systems.

Activity 23

The dangers would include inaccurate data being input, inaccurate processing of data and deliberate breaches of security, such as fraud, hacking and viruses. The ways in which breaches could be prevented include verification and validation checks on data entered, adequate testing of new or amended systems, locks, passwords, user logs, etc.

Activity 24

Your list may have included the following suggestions:

- letter
- memo
- notice
- in-house publication
- manual
- handbook
- computer printout
- fax
- E-mail.

You might decide that it is more appropriate to select an oral communication method such as:

- face-to-face conversation
- meeting
- interview
- telephone conversation
- presentation
- briefing
- training session.

Activity 25

You may have included in your list the fact that the recipient of a written message can read the message when they want to. Written communications can carry complex information and be widely circulated. They provide a permanent record.

Written communication will evoke fewer immediate emotional responses if a difficult message is delivered.

Activity 26

The following are suggestions for appropriate answers:

1 Fax or E-mail
2 Notice
3 Team briefing or presentation
4 Letter or telephone call.

Activity 30

Your interviewer might have made some of the following interview errors. Try not to make any of the common mistakes listed when you are conducting an interview.

- Asking too many inappropriate leading questions
- giving opinions and judgements
- not pursuing points in sufficient depth
- talking too much
- being biased
- not getting at the facts
- not listening to the answers
- not summarizing regularly.

Activity 31

You might have included some of the following reasons in your list:

- *information giving* - to inform people what is happening
- *consultation* - to collect information, views and opinions from people
- *decision-making* - to make a decision which requires input from a group of people
- *problem solving* - to solve a problem which is affecting a group of people
- *to comply with legislation* (e.g. safety meeting) or organizational policy (e.g. management meeting).

Activity 32

You might have included some of the following reasons in your answer:

- they are a good way of exchanging ideas and opinions
- they encourage a good team spirit

- they enable information to be passed on to a group of people
- they can result in better quality decision making and problem solving
- they encourage participation.

There are other valid reasons which you may have included in your list.

Activity 34

The message to be communicated	Speaking and/or writing	
Noxious fumes have escaped into the factory. It must be evacuated immediately	Speaking	You may follow these up with written communication
A member of your workteam is carrying out a task without proper regard for health and safety	Speaking	
You are walking around the factory and notice that a work area is very untidy	Speaking	
You witness an accident in the factory	Written report	
You need to inform 100 of your customers about a changed specification for one of your products	Writing	

Activity 35

You should have listed the following advantages:

1. the receiver of the message can read it when they choose to
2. written communications can carry complex information which would be difficult to describe by speaking
3. they can be widely circulated – sometimes it is important to send the same message to a large number of people
4. if you are delivering a 'difficult' message written communication will give the receiver time to absorb the message before reacting.

We can add to this list:

- written communication creates a permanent record
- the receiver does not have to remember
- can be used as a reference at a later date
- can communicate detail
- ensures that all the recipients have the same message.

Activity 36

Did you remember? Clear, complete, concise, correct and courteous.

Activity 37

Obviously you will have used your own words but the rewritten letter might be something like this.

Dear Mr James

Thank you for your recent letter.

Unfortunately we are having difficulties obtaining a vital component for the product we supply to you. Our suppliers have always been very reliable in the past. I have a meeting with our suppliers this week and I will let you have new delivery dates as soon as possible.

I am very sorry we have not been able to meet our delivery date, as you are a valued customer. I will give you a call later this week to let you know how matters are progressing.

Yours sincerely

David Pratt

David Pratt
Production Supervisor

Activity 39

Details of the communication	Vertical/ upward	Vertical/ downward	Lateral
A briefing meeting for team leaders where your manager briefs you about a new customer		✓	
A meeting of team leaders to discuss common problems			✓
A briefing sheet to all staff from the Managing Director		✓	
A memo from you to your team		✓	

Activity 40

Subject covered at team briefing	Four 'P' classification
Financial results	**Progress**
Organizational mission statement	**Policy**
Time keeping	Depends on the content of the brief; it is likely to be **People**
Dealing with customer complaints	**Points for action**

Activity 41

The first instruction probably makes you feel threatened, under pressure and undervalued. The second instruction should make you feel valued and supported. You understand why the task is important. You know that you can refer to the team leader giving you the instruction if you run into problems. The team leader has treated you with respect and has been courteous.

Activity 42

Of course we all give instructions in our own way and you will have used your own words in the answer. You might have said something like this.

'Dave, Stuart, can I have a quick word with you? I need you to finish tidying up as soon as possible. The 'screw feed hopper' needs cleaning out before we start the next production run which is planned for after lunch. I would like you two to clean the machine, before lunch, because it's essential that the product we are making on the machine on the next run is contamination-free. This product is for a new customer. It is important that we get it right first time. Here is the 'instruction sheet' on cleaning the 'screw feed hopper', make sure that you sign against each task when it has been completed. Is that clear? If you have any questions give me a shout. Don't forget to bring me the completed 'instruction sheet' as soon as the job's finished. Thanks.'

Activity 44

You might have mentioned that you are worried about how the member of staff will react or about creating an unpleasant atmosphere. You might be concerned that you do not know what to say or how to go about giving feedback or you may think that it is a waste of time and that feedback will not improve performance.

Activity 45

You should get the following answers:

Sales (23 per cent), Distribution (5 per cent), Canteen (3 per cent), Wages (2 per cent), Personnel (4 per cent), Research (5 per cent), Administration (6 per cent), Security (2 per cent) and Purchasing (5 per cent).

Activity 46

You should get the following answers:

Sales (24), Distribution (2), Canteen (4), Wages (1), Personnel (4), Research (8), Administration (9), Security (1), Purchasing (4).

Activity 47

Department	All employees		Female employees		Male employees	
	Number	%	Number	%	Number	%
Sales	43	23	24	55	19	45
Distribution	9	5	2	20	7	80
Canteen	6	2	4	67	2	33
Wages	4	3	1	25	3	75
Personnel	8	4	4	50	4	50
Research	10	5	8	80	2	20
Production	85	45	21	25	64	75
Administration	11	6	9	75	2	25
Security	3	2	1	33	2	67
Purchasing	9	5	4	45	5	55
Total	188	100	78	41	110	59

Activity 48

The answers are:

- *arithmetic mean* $= \dfrac{\text{Total of all items}}{\text{Number of items}} = \dfrac{107}{19} = 5.63$
- *median* - 2 2 2 3 3 4 4 5 5 5 6 6 6 6 6 6 6 6 12 12 - the middle item is the tenth, i.e., size 5
- *mode* - the size occurring most frequently is size 6 (eight customers)

You would stock size 6 as being the most popular size. No-one takes a shoe size of 5.6.

Activity 49

You should get 100 for the base year, followed by 117, 108, 125, 133, 125.

Activity 50

You should get 30, 31, 38, 45 and 63.

Activity 51

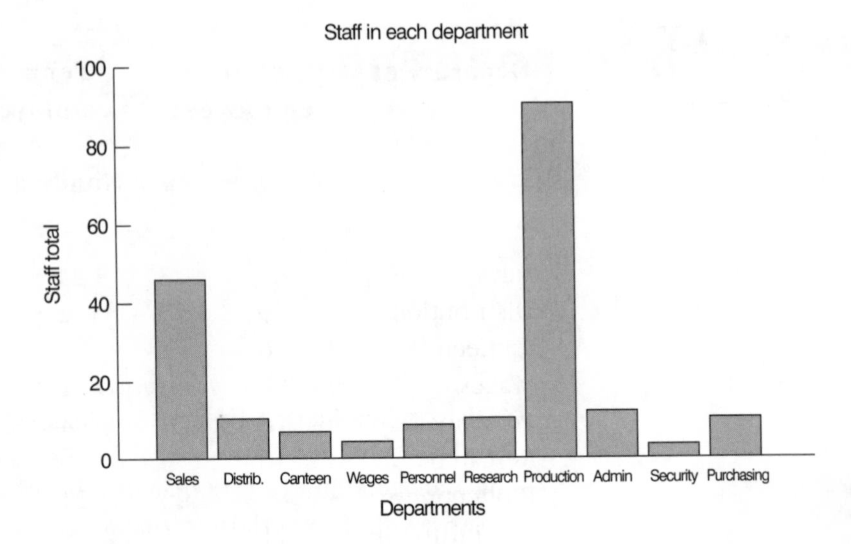

Activity 52

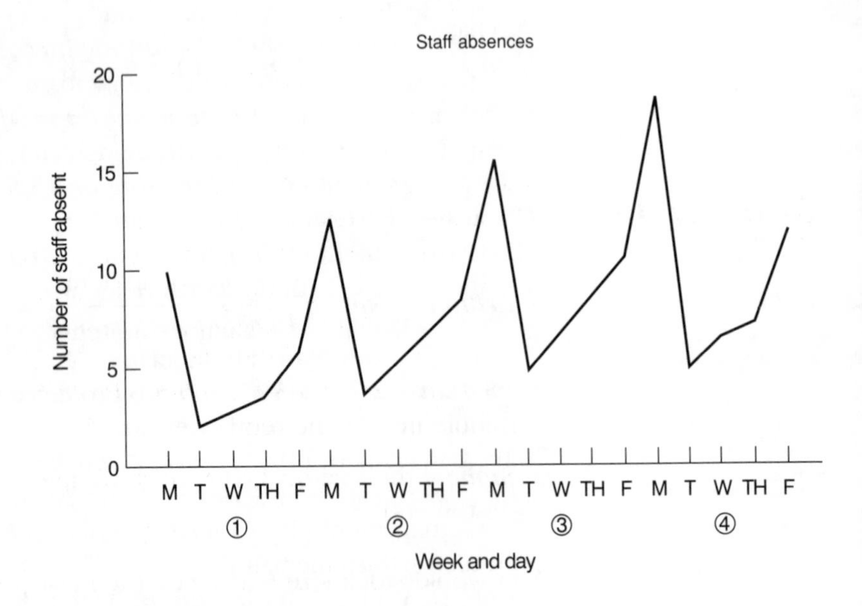

Further reading

ACAS (1997) *Employee Communications and Consultation* (Advisory booklet), Advisory Conciliatory and Arbitration Service

NEBS Management (1997) *Collecting Information* (Super Series 3), Butterworth-Heinemann

NEBS Management (1997) *Communicating in Groups* (Super Series 3), Butterworth-Heinemann

NEBS Management (1997) *Communication in Management* (Super Series 3), Butterworth-Heinemann

NEBS Management (1997) *Information in Management* (Super Series 3), Butterworth-Heinemann

NEBS Management (1997) *Listening and Speaking* (Super Series 3), Butterworth-Heinemann

NEBS Management (1997) *Making and Taking Decisions* (Super Series 3), Butterworth-Heinemann

NEBS Management (1997) *Project and Report Writing* (Super Series 3), Butterworth-Heinemann

NEBS Management (1997) *Solving Problems* (Super Series 3), Butterworth-Heinemann

NEBS Management (1997) *Storing and Retrieving Information* (Super Series 3), Butterworth-Heinemann

NEBS Management (1997) *Writing Effectively* (Super Series 3), Butterworth-Heinemann

Wilson, D. A. (1997) *Managing Information* (second edition), Butterworth-Heinemann

Index